T0363990

ShapingourEnvironments

LEVEL3GEOGRAPHY

Interacting Natural Processes in a Coastal Environment

Matthew Haines

Interacting Natural Processes in a Coastal Environment
1st Edition
Matthew Haines

Cover design: Cheryl Smith, Macarn Design
Text designer: Cheryl Smith, Macarn Design
Production controller: Siew Han Ong

Any URLs contained in this publication were checked for currency during the production process. Note, however, that the publisher cannot vouch for the ongoing currency of URLs.

Acknowledgements
Shutterstock for the images on the front cover and pages 4, 5, 6, 14, 19, 22, 23, 27, 30, 32, 35, 36 (top), 42, 45, 46, 47, 48, 49, 50, 51, 54, 59, 60, 65, 70, 75, 76, 85, 86 (lower), 87.
iStock for images on back cover and pages 36 (lower), 72, 77, 81, 86 (top).
NASA for satellite image on page 40.
Alexander Turnbull Library, Wellington, New Zealand for images on page 72 (top left and right) 1951 and 1957 aerial photographs taken by Whites Aviation Ltd.

© 2020 Cengage Learning Australia Pty Limited

Copyright Notice
Copyright: This book is not a photocopiable master. This book is sold on the basis that it will be used by an individual teacher or student. No part of the publication may be copied, stored or communicated in any form by any means (paper or digital), including recording or storing in an electronic retrieval system, without the written permission of the publisher. Copying permitted under licence to Copyright Licensing New Zealand **does not** extend to any copying from this book.

For product information and technology assistance,
in Australia call **1300 790 853**;
in New Zealand call **0800 449 725**

For permission to use material from this text or product, please email
aust.permissions@cengage.com

National Library of New Zealand Cataloguing-in-Publication Data
A catalogue record for this book is available from the National Library of New Zealand

978 0 17 044690 7

Cengage Learning Australia
Level 7, 80 Dorcas Street
South Melbourne, Victoria, Australia 3205

Cengage Learning New Zealand
Unit 4B Rosedale Office Park
331 Rosedale Road, Albany, North Shore 0632, NZ

For learning solutions, visit **cengage.co.nz**

Printed in China by 1010 Printing International Limited.
3 4 5 6 7 24

Contents

1 Introduction to coasts

What is a coast?

In simple terms, a coast is where the land meets the sea.

However, coasts are more complex than that. A coast is an environment on the Earth's surface where three key elements interact: the ocean, the land and the atmosphere. The interactions between these elements can be dramatic and result in coastlines being dynamic and ever-changing. Furthermore, many coastlines around the world are used by humans, as places to live, work and play. As a result, coasts have become one of the most studied types of environments within the discipline of Geography.

Coastal Geography is an important branch of Geography. It is an area of speciality at universities in New Zealand and around the world. Coastal Geography can lead to a number of exciting career opportunities, including coastal management, coastal ecology, tourism and environmental engineering.

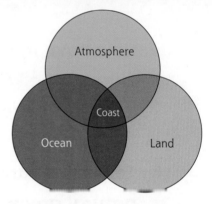

    PHOTOCOPYING OF THIS PAGE IS RESTRICTED UNDER LAW.   ISBN: 9780170446907

What do coasts look like?

Look at the two pictures below. What do you see? Coastlines look vastly different all around the world. At a most basic level, coastlines may be cliffed coasts or rocky coasts, or may comprise sandy or stony beaches.

Another important coastal feature is headlands (see below). These are rocky outcrops of erosion-resistant rock that protrude out into the sea. The areas between these headlands comprise softer, more vulnerable rock and have been eroded back by the waves faster than the headlands. Curved beaches, known as bays, may form between two headlands.

Bay

Headlands

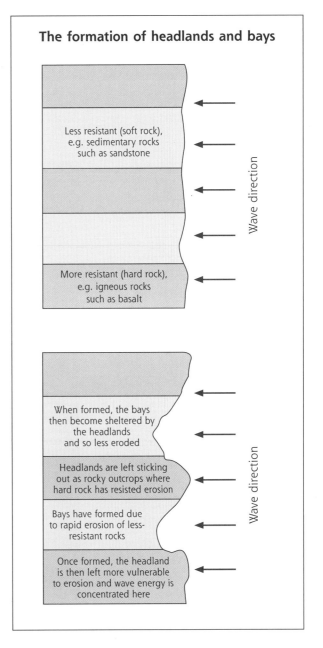

The formation of headlands and bays

Less resistant (soft rock), e.g. sedimentary rocks such as sandstone

More resistant (hard rock), e.g. igneous rocks such as basalt

Wave direction

When formed, the bays then become sheltered by the headlands and so less eroded

Headlands are left sticking out as rocky outcrops where hard rock has resisted erosion

Bays have formed due to rapid erosion of less-resistant rocks

Once formed, the headland is then left more vulnerable to erosion and wave energy is concentrated here

Wave direction

Beaches on coasts can be divided into key zones:

1 Backshore – this is the area at the top of a beach and is not touched by waves (except during storms and/or very high tides). This area may also include cliffs or sand dunes.
2 Foreshore – area between high- and low-tide marks. Is covered by the sea during high tide and exposed during low tide.
3 Inshore – sometimes called the surf zone, this is the area where the sea floor is shallow enough to cause waves to break.
4 Offshore – this is the area out to sea beyond where the waves break (too deep for waves to break).

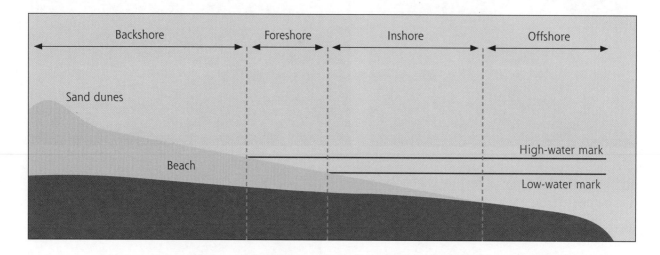

It is also possible to categorise coasts according to the amount of energy in marine processes that are occurring.

Low-energy coasts:
- stretches of the coastline where **waves are not powerful**
- often the rate of deposition exceeds the rate of erosion
- landforms include beaches and spits
- these are called **constructive coastlines**.

High-energy coasts:
- stretches of the coastline where **waves are powerful** for a significant part of the year
- often the rate of erosion exceeds the rate of deposition
- landforms include headlands, cliffs and wave-cut platforms
- these are called **destructive coastlines**.

 PHOTOCOPYING OF THIS PAGE IS RESTRICTED UNDER LAW. ISBN: 9780170446907

Tasks

1 In your own words, define the term 'coast'.

_____

_____

2 Do all coasts look the same? Discuss.

_____

_____

_____

_____

3 Draw a diagram to illustrate the four different zones of a beach. Add annotations that describe these different zones. (Hint: diagrams should be clear, neat and have a title.)

4 What are constructive and destructive coastlines? Describe what makes them different.

_____

_____

_____

_____

_____

_____

2

Introduction to coastal processes

Coastal processes are actions, movements or transfers of energy that happen in the natural environment. They happen at different rates and scales over time and space and play a dramatic role in shaping coastal environments. At coastlines these processes are mostly driven by energy from the sun, except in the case of tectonic processes where the energy comes from the Earth's core. Marine (wave) processes, aeolian (wind) processes, biological and sub-aerial processes, as well as tectonic processes are some of the most significant forces that shape coasts. It is important to note that these processes interact (work together) to produce noticeable results on the shape of the coastal environment.

Shaping of the coastal environment – marine (wave) processes

Waves are created by the action of wind blowing over the surface of the sea. Wave size and energy depends on:

- **wind strength**
- **wind duration** (how long the wind is blowing)
- the **fetch** of the wave (the maximum distance of open sea a wave can travel before it hits land).

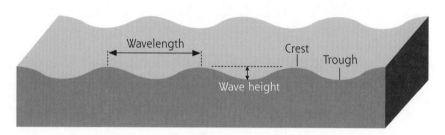

The highest part of a wave is the **crest** and the lowest point is the **trough**. The difference between crest and trough is the **wave height**. The distance between one crest and the next is the **wavelength**.

Large slow-moving waves out in the open ocean are referred to as **swells**. Swells are defined as mature undulations of water in the open ocean (out at sea, not near coastlines) after wave energy has moved on from where it originated. Like other waves, swells can range in size from very small to enormous. A key thing to understand about swells is that while it looks like the water is moving forward, only a small amount of water is actually moving. It is not water but the wave's energy that is moving. If you imagine you were sitting in a boat out at sea in swells, you would not be significantly moving in any direction – simply bobbing around in a circular motion (see the left-hand side of the 'Breaking waves' diagram on page 9).

As the wave travels towards the shore and the depth decreases, the diameter of these circular patterns also decreases. When the diameter decreases, the patterns become oval-shaped and the entire wave's speed slows. Because waves move in groups, they continue arriving behind the first and all of the waves are forced closer together since they are now moving more slowly. They then grow in height and steepness. When the waves become too high relative to the water's depth, the wave's stability is undermined and the entire wave topples onto the beach forming a **breaker** (see diagram on page 9). When waves break on a beach, some of the water and energy is reflected back into the sea, some infiltrates into the sediment (sand or stones) on the beach, and other water rushes up the face of the beach. This foamy water rushing up the beach is called **swash**. The swash eventually slows down because of gravity and the slope of the beach and then runs back down the beach, and is called **backwash**.

 PHOTOCOPYING OF THIS PAGE IS RESTRICTED UNDER LAW. ISBN: 9780170446907

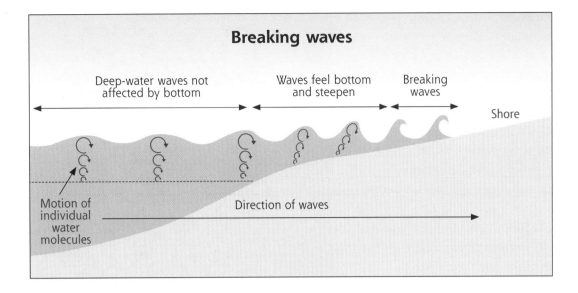

Breaking waves

Deep-water waves not affected by bottom

Waves feel bottom and steepen

Breaking waves

Shore

Motion of individual water molecules

Direction of waves

Constructive waves and destructive waves

There are two types of wave: **constructive waves** and **destructive waves**.

Constructive waves have limited energy. They have a strong swash that transports material up the beach increasing the amount of beach material and creating a steeper-sloped and shorter beach. Constructive waves appear lower in height and are less frequent (about six to eight waves per minute).

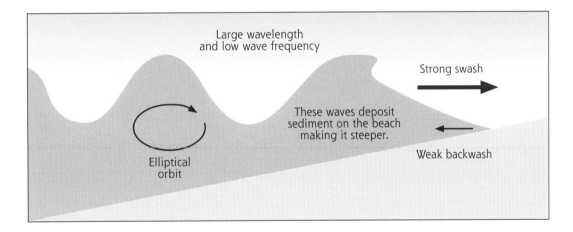

Large wavelength and low wave frequency

Strong swash

These waves deposit sediment on the beach making it steeper.

Weak backwash

Elliptical orbit

Destructive waves have much more energy. They have a strong backwash that transports material back down the beach reducing the amount of beach material and creating a big flat wide beach. Destructive waves appear to be higher and more frequent (about 12–14 waves per minute).

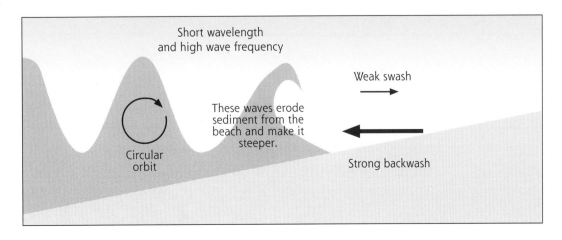

Short wavelength and high wave frequency

Weak swash

These waves erode sediment from the beach and make it steeper.

Strong backwash

Circular orbit

Wave refraction

The direction in which a wave moves may be altered by the shape of the coastline. Waves travel faster in deeper water. If a wave approaches the coast at an angle, the side nearer the coast, in shallower water, loses more energy to friction so slows down. This causes the wave to refract (change direction).

The direction of the waves is affected by features such as islands, bays and headlands. Refraction around a headland can result in erosional formations on each side of the headland (such as Otakamiro Headland at Muriwai Beach).

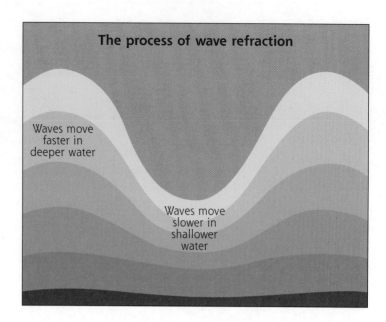

The process of wave refraction

Waves move faster in deeper water

Waves move slower in shallower water

⊙ Tasks

1 In your own words, define the term 'wave'.

2 Explain how ocean waves are formed. What factors can influence how big waves become?

3 Draw a diagram to illustrate the key characteristics of ocean waves (wave length, wave height, crest, trough).

   PHOTOCOPYING OF THIS PAGE IS RESTRICTED UNDER LAW.   ISBN: 9780170446907

4 Explain in detail the key differences between waves in the open ocean and waves near the coast. Explain why waves at the coast 'break'.

_____

_____

_____

_____

_____

_____

_____

_____

5 Define the following terms.

Swash: _____

Backwash: _____

6 Annotate the blank diagrams below to illustrate the key differences between constructive waves and destructive waves. Add annotations that provide additional detail.

7 With the aid of a diagram, explain wave refraction.

_____

_____

_____

_____

_____

_____

_____

_____

How can waves shape coastlines? Erosion

Marine erosion

What is erosion? Erosion is the gradual destruction and removal of sediment, soil or rock from a particular location. There are a number of forces of nature that can erode the land including wind (**aeolian processes**), rivers (**fluvial processes**) and glaciers (**glacial processes**). Coastlines can be eroded by these three forces – coastlines are windy, coastlines are dotted with the mouths of rivers and coastlines may exist in polar regions where glaciers are prominent.

However, the most obvious force of nature that will erode coastlines and modify them are waves. Waves are one of the most powerful forces on planet Earth, and the constant day-to-day, relentless hammering away of them on coastlines is highly destructive. This is the main reason why coastlines change so much.

Beaches are made of unconsolidated (loose) sediment, such as sand or pebbles, and destructive waves can plunge onto them and drag away large amounts of this sediment over a period of time. During winter many beaches lose thousands of tonnes of sediment due to the high rates of wave erosion. This force of the water dislodging sediment and taking it away is called **hydraulic action**.

 PHOTOCOPYING OF THIS PAGE IS RESTRICTED UNDER LAW. ISBN: 9780170446907

Cliffs and headlands are made of consolidated material and are much more resistant to erosion than beaches – they erode much more slowly. Waves can erode cliffs and headlands in different ways:

- **Hydraulic action** – This is erosion that is the result of the sheer force of water moving past a surface. Waves can dislodge loose rocks from a cliff and carry them away. **Wave pounding** is a type of hydraulic action where rock is weakened by the constant hammering of heavy waves.
- **Wave hydraulic pressure** – This is actually a type of hydraulic action. On cliff faces, air present in joints and fissures (small cracks in the rock) is trapped and compressed by the pressure of incoming sea-water. Over a period of time, this increase in air pressure weakens and breaks off the rock. The rate of hydraulic pressure is high on coasts where waves are powerful and the coastline is made up of a densely jointed rock.
- **Abrasion (corrasion)** – Sand, shingle and boulders, carried by the sea, rub against the surface of cliffs and scour (wear) them down. It is the fastest and most effective form of coastal erosion. It is particularly powerful in coastlines where waves are transporting large amounts of sediment.
- **Attrition** – This is a type of erosion which affects already broken off rocks, pebbles and boulders that are rolling around in the surf. The movement of waves makes rocks and pebbles crash together, so that sharp edges are broken down, and particles become smaller and more rounded. It affects boulders and stones that have already been eroded from the coast.
- **Solution (corrosion)** – This is a type of chemical erosion where acids present in water can erode cliffs.

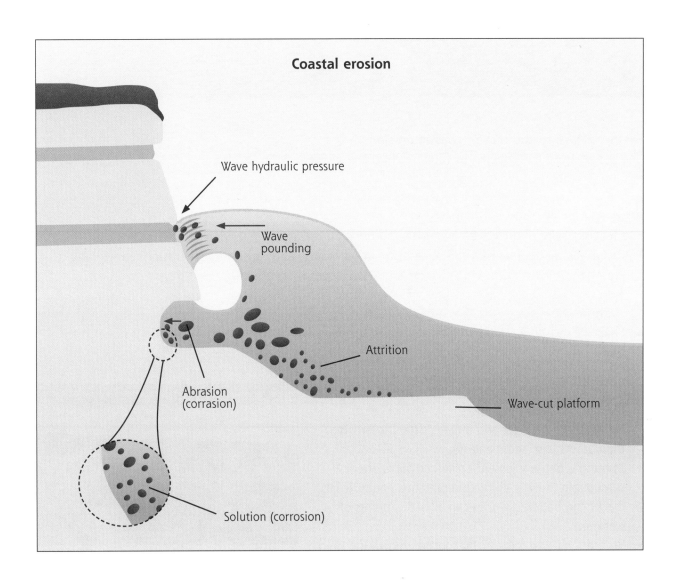

Coastal erosion

Wave hydraulic pressure

Wave pounding

Attrition

Abrasion (corrasion)

Wave-cut platform

Solution (corrosion)

Factors affecting the rate of coastal erosion

The rate of erosion is affected by the force of the waves (erosivity) and the resistance of the coast to erosion (erodibility).

What determines the force of the waves?

- Breaking point of the wave – when a wave breaks it releases a great deal of energy. A wave that breaks at the foot of a cliff releases the most energy and causes fastest erosion, particularly corrasion. A wave that breaks offshore will have lost most of its energy as it travels up a beach.
- Type of wave – steep destructive waves have more energy, and power to erode, than shallow constructive waves.
- Shape of coastline – refraction makes waves stronger and more erosive on headlands rather than bays.

What determines the resistance of the coast to erosion?

- Mechanical strength of rocks – some rocks (e.g. igneous) are stronger and more resistant to erosion than others (e.g. sedimentary).
- Jointing – densely jointed or faulted rocks are susceptible to hydraulic action/pressure. Faults, joints, cracks and bedding planes can all act as points of weakness.
- Vegetation – the foliage and roots of vegetation bind soil and rocks together and reduce the rate of erosion.
- Human protection – in many locations, physical structures (e.g. sea walls) have been installed to absorb the energy of waves and so reduce the rate of erosion.

PHOTOCOPYING OF THIS PAGE IS RESTRICTED UNDER LAW.   ISBN: 9780170446907

Tasks

1　In your own words, define the term 'erosion'.

_____

_____

_____

2　Provide your own geographical definition for each of these types of marine erosion.
- Wave pounding

_____

_____

_____

_____

- Wave hydraulic pressure

_____

_____

_____

_____

- Abrasion

_____

_____

_____

_____

- Attrition

_____

_____

_____

_____

- Solution

_____

_____

_____

_____

3 Draw annotated diagrams to illustrate and explain these different types of marine erosion.

Wave pounding

Wave hydraulic pressure

Abrasion

    PHOTOCOPYING OF THIS PAGE IS RESTRICTED UNDER LAW.   ISBN: 9780170446907

<table>
<tr><td>**Attrition**</td></tr>
<tr><td>

</td></tr>
</table>

<table>
<tr><td>**Solution**</td></tr>
<tr><td>

</td></tr>
</table>

4 Explain the difference between 'erosivity' and 'erodibility'.

• erosivity

_____

_____

_____

_____

• erodibility

_____

_____

_____

_____

5 Discuss the factors and circumstances that influence the erosivity and erodibility of a coastline.

_____

_____

_____

_____

_____

_____

_____

_____

_____

_____

_____

_____

_____

_____

_____

_____

_____

_____

_____

_____

_____

_____

_____

_____

_____

_____

_____

_____

_____

_____

_____

_____

_____

_____

_____

_____

PHOTOCOPYING OF THIS PAGE IS RESTRICTED UNDER LAW.   ISBN: 9780170446907

Shaping of the coastal environment – sub-aerial processes

Sub-aerial processes are those processes that operate at the coast but do not involve direct contact with the sea. It is a type of weathering – the loosening and weakening of rock.

- **Salt crystallisation** Sea spray enters cracks. Later the water evaporates to leave crystals of salt. Further evaporation enlarges the crystals. The growing crystal exerts force on the rock. The rate of salt weathering is most rapid in well-jointed rocks (porous rocks with lots of cracks and fissures).
- **Mass movement** This is particularly active at the coast because undercutting of rocks by the sea makes them unstable. Rockfalls occur when the waves undercut the cliffs and weathering loosens pieces of rocks on the cliff face. Rockfalls are most common on cliffed coastlines with resistant rocks such as Manukau Breccia and results in cliff retreat.

Interaction of sub-aerial processes and marine erosion

In natural environments processes interact. This means that as one process operates it can affect the way another process operates. It can be a one-way interaction or a two-way interaction.

As a geographer, the easiest way to identify and discuss an interaction is to look at whether a process speeds up, slows down or stops another process.

Can sub-aerial processes interact with marine erosion? Definitely, yes. And it may be a two-way interaction. Sub-aerial processes, such as salt crystallisation, will weaken the structure of rock on a cliff – this will in turn increase rates of marine erosion such as hydraulic action and abrasion. Therefore, the higher the rate of salt crystallisation, the higher the rate of marine erosion.

Conversely, high rates of marine erosion will increase rates of sub-aerial processes. For instance, if waves are powerful, they will undercut a cliff more rapidly. This will, in turn, weaken the cliff considerably making it very unstable. As a result, the sub-aerial process of mass movement will occur at a much faster rate than if wave erosion was not as destructive.

Tasks

1 Fully explain sub-aerial weathering, referring to at least two different types.

_____

_____

_____

_____

_____

_____

_____

_____

_____

_____

_____

_____

_____

_____

_____

_____

_____

_____

2 Explain how sub-aerial weathering differs from marine erosion.

_____

_____

_____

_____

3 In a general sense, explain what is meant by 'interacting processes'.

_____

_____

_____

_____

_____

_____

_____

PHOTOCOPYING OF THIS PAGE IS RESTRICTED UNDER LAW.   ISBN: 9780170446907

4 In the empty diagrams below, explain interactions between the processes of sub-aerial weathering and marine erosion.

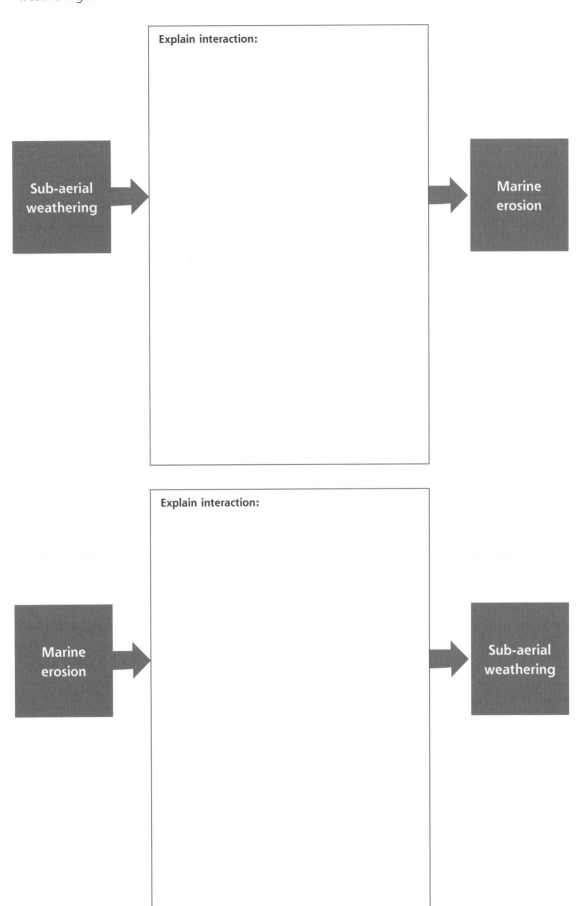

Explain interaction:

Sub-aerial weathering → Marine erosion

Explain interaction:

Marine erosion → Sub-aerial weathering

Erosional features – wave-cut notches and shore platforms

When high and steep waves break at the foot of a cliff, they concentrate their erosive capabilities into only a small area of the rock face. This concentration eventually leads to the cliff being undercut, forming a wave-cut notch. Continued activity at this point increases the stress on the cliff and after some time, it collapses. This causes the cliff to retreat and when the overhang is undercut a platform forms. The platform continues to grow and, as it does, the waves break further out to sea and have to travel across more of the platform before reaching the cliff. This leads to greater dissipation of wave energy, which reduces the rate of erosion on the headland, thereby slowing down the growth of the platform. Hydraulic action and abrasion, interacting with salt crystallisation weathering, lead to the formation of wave-cut notches and platforms.

Wave-cut notch

Shore platform

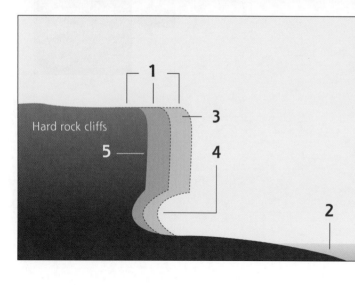

Hard rock cliffs

1 Weathering weakens the top of the cliff.
2 The sea attacks the base of the cliff, forming a wave-cut notch.
3 The notch increases in size, causing the cliff to retreat.
4 The backwash carries the rubble towards the sea, forming a wave cut platform.
5 The process repeats and the cliff continues to retreat.

   PHOTOCOPYING OF THIS PAGE IS RESTRICTED UNDER LAW.   ISBN: 9780170446907

Erosional landform features – caves, arches and stacks

Weathering and *erosion* can create caves, arches, stacks and stumps along a headland. Marine erosion can be highly concentrated along the sides of a headland due to interaction with the process of wave refraction.

- **Caves** occur when waves force their way into cracks (known as joints or fissures) in the cliff face. The water contains sand and other materials that grind away at the rock until the cracks become a cave. *Hydraulic pressure* is the predominant process.
- If the cave is formed in a headland, it may eventually break through to the other side forming an **arch**.
- The arch will gradually become bigger until it can no longer support the top of the arch. When the arch **collapses**, it leaves the headland on one side and a **stack** (a tall column of rock) on the other.
- The stack will be attacked at the base in the same way that a wave-cut notch is formed. This weakens the structure and it will eventually **collapse** to form a **stump**.

Headland features resulting from erosion

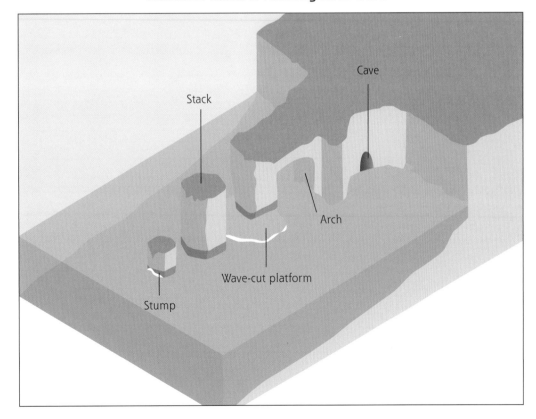

ISBN: 9780170446907   PHOTOCOPYING OF THIS PAGE IS RESTRICTED UNDER LAW.

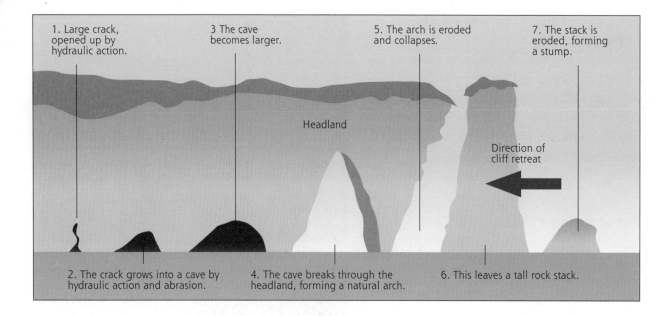

1. Large crack, opened up by hydraulic action.

3 The cave becomes larger.

5. The arch is eroded and collapses.

7. The stack is eroded, forming a stump.

Headland

Direction of cliff retreat

2. The crack grows into a cave by hydraulic action and abrasion.

4. The cave breaks through the headland, forming a natural arch.

6. This leaves a tall rock stack.

⦿ Tasks

1 Write a brief description of the following types of erosional coastal landforms, saying what they look like:

- Wave-cut notch

_____

_____

_____

- Shore platform

_____

_____

_____

- Arch

_____

_____

_____

- Stack

_____

_____

_____

- Stump

_____

_____

_____

 PHOTOCOPYING OF THIS PAGE IS RESTRICTED UNDER LAW.   ISBN: 9780170446907

2 Explain how marine erosion leads to cliff retreat and forms a shore platform. You may include a diagram to assist in your answer.

_____

_____

_____

_____

_____

_____

_____

_____

_____

_____

_____

_____

_____

3 Draw a detailed annotated diagram (or series of detailed annotated diagrams) that fully explains how marine erosion interacting with other natural processes results in the formation of arches, stacks and stumps.

Shaping of the coastal environment – marine transportation processes

Marine transportation is extremely powerful at transporting (moving) large quantities of eroded sediment. This material can be moved (transported) by waves and currents. Wave transportation and currents can carry sand and other material towards the land. Destructive waves and rip currents (strong currents that move water away from a beach) carry large amounts of sediment away from the land to the offshore zone.

There are a number of different ways that energy in water currents and waves can move sediment, as shown below.

Process	Description
Traction	Pebbles and larger sediment are rolled along the sea bed.
Saltation	Sediment such as sand is bounced along the sea bed in a skipping motion. This happens because the currents cannot keep the larger and heavier sediment afloat for long periods.
Suspension	Small particles (e.g. silts and clays) float in water currents. This can make the water look cloudy. Currents will pick up large amounts of sediment in suspension during a storm, when strong winds generate high-energy waves.
Solution	Certain compounds are dissolved into the water and carried away.

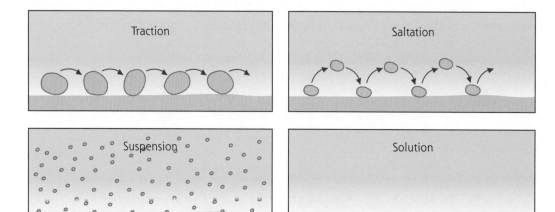

Longshore (littoral) transportation

At coastlines, large quantities of sediment are moved along the coast by the force of the waves. This 'longshore transportation' can be split up into two main types: **littoral drift** and **longshore drift**.

- **Littoral drift** Currents can move from one end of a beach to another (known as the 'littoral current'). This current happens in the surf zone and is due to waves approaching from an angle.

- **Longshore drift** Waves can approach the coast at an angle because of the direction of the prevailing wind. The swash of the waves carries material up the beach at an angle. The backwash then flows back to the sea in a straight line at 90°. Continual swash and backwash transports material sideways along the coast. This movement of material is called longshore drift and occurs in a zigzag motion.

 PHOTOCOPYING OF THIS PAGE IS RESTRICTED UNDER LAW. ISBN: 9780170446907

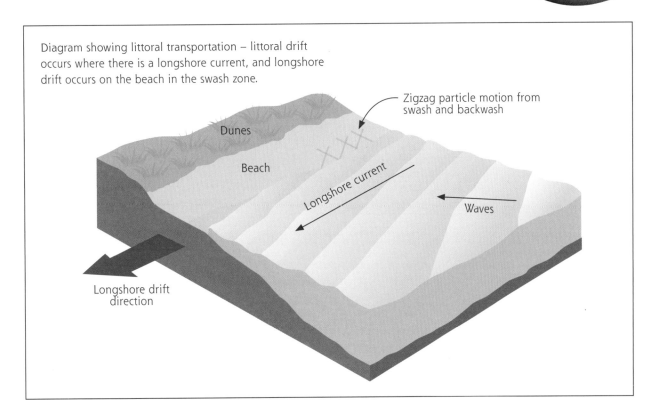

Diagram showing littoral transportation – littoral drift occurs where there is a longshore current, and longshore drift occurs on the beach in the swash zone.

Zigzag particle motion from swash and backwash

Dunes

Beach

Longshore current

Waves

Longshore drift direction

The photo shows sediment building up on one side of the man-made structure due to longshore drift.

ISBN: 9780170446907   PHOTOCOPYING OF THIS PAGE IS RESTRICTED UNDER LAW.

Tasks

1 Write a description of each of the following types of marine transportation, and include a simple diagram.

Traction:

_____

_____

_____

_____

Saltation:

_____

_____

_____

_____

Suspension:

_____

_____

_____

_____

Solution:

_____

_____

_____

_____

2 Fully explain the difference between littoral drift and longshore drift.

_____

_____

_____

_____

_____

_____

_____

_____

_____

_____

_____

    PHOTOCOPYING OF THIS PAGE IS RESTRICTED UNDER LAW.   ISBN: 9780170446907

3 Draw a diagram that illustrates the different processes of littoral transportation (longshore drift *and* littoral drift). Add detailed annotations explaining these processes.

Shaping of the coastal environment – marine deposition processes

Waves that carry sediment are able to deposit it at points along a coast. Waves that deposit sediment are gently sloped, low-energy waves. Their swash pushes sand onto a beach and they have very little backwash (most of their energy dissipates into the sediment on the beach).

A beach is a depositional feature. It is completely made up of eroded material that waves have deposited. Beaches sit mostly between the high-tide and low-tide mark of a coastline. For much of the day they are covered by the water of the high tide.

Beaches are dynamic and forever-changing, especially when looking at seasonal patterns. This is a direct result of the interaction between climatic processes and wave processes. In the summer, high air-pressure (anti-cyclonic) conditions in the atmosphere result in lower wind speeds. Less wind energy is transferred into ocean waves in summer and therefore the waves are more gentle and constructive. These constructive waves push sediment onto the beach and deposit it to form a steeper-sloped beach and a berm (a ridge of sand that runs along the beach at the high-tide mark). It is as if the waves during summer act like little bulldozers, slowly pushing more and more sand onto the beach.

During winter, climatic conditions tend to be more unstable. Low air-pressure systems result in stronger wind patterns and there is a higher occurrence of storms and cyclones. This weather interacts with marine processes, making waves bigger, stronger and steeper. Waves in winter are more likely to be destructive, and with a higher ratio of backwash to swash they erode sand off the beach. Large quantities of sediment are taken off the beach and carried out to the offshore zone, forming large bars of sand on the sea bed off the coast. In fact, these sand bars act as a natural coastal defence, causing waves to break further out to sea, reducing the erosivity of them when they reach the beach. Winter beaches are flat and without a berm. At high tide, waves can reach all the way up the beach and undercut the front of sand dunes.

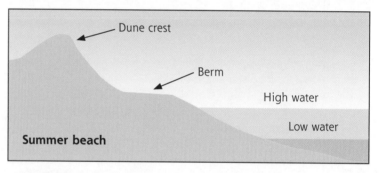

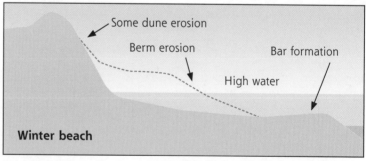

PHOTOCOPYING OF THIS PAGE IS RESTRICTED UNDER LAW.   ISBN: 9780170446907

Tasks

1 Explain how beaches can vary between summer and winter.

_____

_____

_____

_____

_____

_____

_____

_____

_____

_____

_____

_____

_____

_____

_____

_____

2 How is this the result of climatic processes interacting with marine processes, in different ways and in different seasons?

_____

_____

_____

_____

_____

_____

_____

_____

_____

_____

_____

_____

_____

_____

_____

Aeolian processes – how can wind shape coastal environments?

Aeolian processes also play an important role in shaping coastal environments. Like waves, wind can erode, transport and deposit. On cliffs at coasts, wind can sculpt rock, eroding small particles of sediment and transporting them away. However, this is very slow and marine and sub-aerial processes definitely play a much more significant role.

However at beaches, where unconsolidated sediment is present, wind is highly effective at shaping distinct landforms. The most iconic beach landform created by aeolian processes at coasts are sand dunes.

At the most basic level, dunes are simply piles of sand. Sand dunes are present on shorelines where fine sediment is transported landward by a combination of wind and waves, and stabilised with vegetation such as spinifex. Primary dunes (or foredunes) are situated nearest to the ocean and are affected most significantly by waves and salt spray. Secondary dunes (or rear dunes) are located further inland and are not often directly exposed to marine influences.

Dunes undoubtedly are one of the most well-known features of sandy beaches, yet they are also not properly understood in many societies and communities around the world – therefore in many instances they are poorly managed. Dunes can help protect settlements along coastlines from the erosion associated with cyclones and tsunamis, however human activities have had severe impacts on coastal sand dunes.

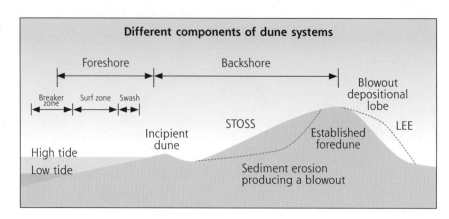

Different components of dune systems

PHOTOCOPYING OF THIS PAGE IS RESTRICTED UNDER LAW.
ISBN: 9780170446907

How are dunes formed and shaped by aeolian processes?

Dunes are formed by a number of interacting natural processes that operate at the coast. While aeolian processes are most important, marine processes begin the sequence of dune formation (see below).

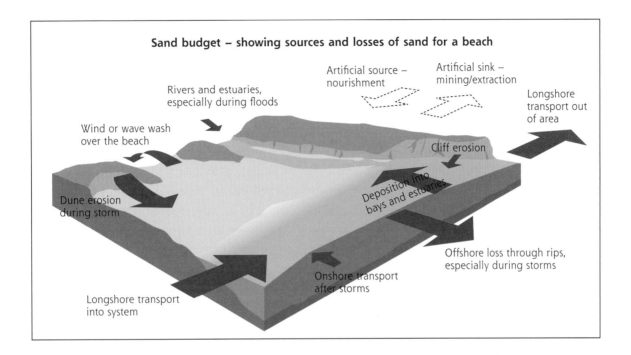

Sediment is transported into a coastal environment, primarily by longshore currents and longshore drift, and from wave transportation from the offshore zone (including from sand bars). Waves then deposit sand, creating and building up a beach. This supplies the sediment required by aeolian processes to form dunes (an important interaction between wave and wind processes – the higher the rate of wave deposition, the higher the rate of aeolian erosion and transportation).

Wind erodes sediment from the beach. This is sometimes called 'deflation'. For deflation to be very effective and contribute towards dune formation, there need to be several key environmental conditions in place:

- Prevailing wind (the most common wind direction) needs to be blowing 'onshore' (from ocean onto the land).
- Wind velocity needs to be fast enough to have the energy required to actually erode sand. This speed is known as the 'fluid threshold velocity' and varies from coast to coast depending on the size and weight of sediment present.
- If the tide is at its lowest, large amounts of sediment will be exposed to the wind and able to be deflated.
- Solar heating of beach sediment exposed at low tide will dry it out and make it friable (loose and not clumped together as when wet). This will make it much easier for wind to erode.

Once wind has eroded particles of sediment off the coast, sediment is then transported. Aeolian erosion and aeolian transportation are interlinked and interrelated – they interact with each other. The more sediment that is eroded, the higher the rates of aeolian transportation. For sediment to be moved distances, the wind needs to maintain its speed so there is enough energy. This usually occurs on beaches as they are very exposed, with little vegetation and the occasional piece of driftwood or seaweed.

There are three main ways that sediment can be transported up a coastline by wind:

1 **Suspension** The finer sand particles are moved by the wind, high in the air. Smaller sediment particles are less affected by gravity and therefore can travel long distances before they land on earth again. This form of aeolian transportation only affects sand grains when wind speeds are very high.

2 **Saltation** This is the most common form of aeolian transportation at coastlines. When the wind hits the ground, it causes turbulence, disturbing the sand particles. If the wind has enough velocity, it will cause the particles to start moving (initially just along the ground). As the sand moves, it hits other grains, which causes them to bounce up in the air. The wind then picks these airborne particles up and carries them. Gravity causes them to fall back down. If sand lands on a hard surface (e.g. rock), the sand particle will bounce off again, being carried further. If it lands on a sandy surface, it will cause other particles to be disturbed, bounce up and they too will be carried, thus starting off a chain reaction.

3 **Surface creep** The larger particles are too heavy to be picked up and carried by the wind so instead, they move along the ground. When they become dislodged by the falling ones, they roll along the ground. Through this process, they are not only moved but by moving against other particles, they erode into smaller particles, which can be moved by saltation or suspension.

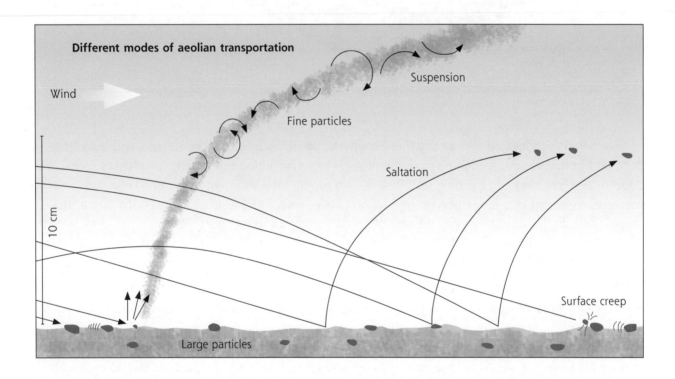

Sand is transported up the beach by these processes from the foreshore to the backshore. If dunes are already present, sediment will travel up the dunes until they are deposited. Aeolian deposition will occur when wind speeds drop low enough for entrained sediment to fall to the ground and stay there. There are several factors that may cause sediment to be deposited. One of these is that wind speed drops on the leeward side of the dune, in the dune slack, because it is sheltered. This will cause an accretion (accumulation) of sediment on this side of the dune.

Another factor very important in causing aeolian deposition is dune vegetation. Dunes are difficult environments for plants to grow on – they are extremely windy and very dry (xerophytic). Only very hardy plants will grow here, mainly short tough grasses with very long roots. Fronds of these grasses will interrupt wind currents at levels near the ground enough for sediment to be deposited around these small plants. As the grass is buried by deposited sand, it will grow outwards contributing to further deposition and further dune accretion. For this reason, dune grasses play a massive role in the shaping of dunes by aeolian processes (another example of a geographic interaction). Dunes play an important role in New Zealand at protecting people from storms, flooding, king tides and climate change. Consequently, coastal managers value dune grasses, often embarking on extensive planting projects and placing rope fences and signage to keep people off the plants.

New foredunes begin their life known as 'embryo dunes'. These are small patches of vegetation found on a beach. Over time these small embryos will collect sand and build up into bigger and bigger dunes, until a foredune system is present. It is extremely difficult for embryo dunes to establish themselves and develop into full-sized dunes because the environment is so dry, saline and windy.

Sometimes dunes themselves can be severely eroded by the wind. Where this happens they are called 'blowouts'. Blowouts commonly occur in the foredune or within and on older vegetated dunes. Blowouts are erosional dune landforms that are typically trough-, bowl- or saucer-shaped depressions. The breaching may occur naturally after erosion of the foredune by storm waves or by funnelling of winds through saddles (lows) in the crest of a high foredune, or wherever there is a reduction in the vegetation cover, thus decreasing the local roughness and increasing the potential for sediment deflation. Over time, there has been an increasing disturbance of dune vegetation associated with human activities such as vegetation clearance for farming and grazing, trampling by farm animals and/ or feral and exotic animals, and urban development. Significant disturbance to the vegetation has also been stimulated by increased recreational activities such as trail bikes, dune buggies, and four-wheel-drive vehicles, as well as by people walking across the dunes. At exposed sites even a few people occasionally walking across a foredune may disturb the vegetation sufficiently to initiate blowouts.

ISBN: 9780170446907 PHOTOCOPYING OF THIS PAGE IS RESTRICTED UNDER LAW.

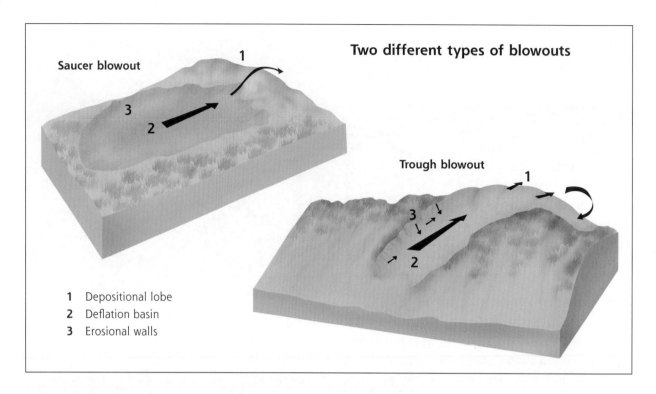

Two different types of blowouts

Saucer blowout

Trough blowout

1 Depositional lobe
2 Deflation basin
3 Erosional walls

Continued transport of sand through blowouts often results in the development of parabolic dunes. These consist of an actively advancing nose and depositional lobe with two trailing arms that enclose a deflation basin. This produces a characteristic U-shaped (i.e. parabolic) or V-shaped dune.

Both images show dunes in New Zealand being affected by blowouts.

 PHOTOCOPYING OF THIS PAGE IS RESTRICTED UNDER LAW. ISBN: 9780170446907

Tasks

1 What are sand dunes?

_____

_____

_____

_____

_____

_____

2 For dune accretion (build-up) to occur, sediment must first be eroded from the beach by wind. In the diagrams below, describe four different conditions that are required for aeolian erosion and dune accretion to occur.

Condition 1	Condition 2

Condition 3	Condition 4

ISBN: 9780170446907   PHOTOCOPYING OF THIS PAGE IS RESTRICTED UNDER LAW.

3 Provide your own geographical definition for the following modes of aeolian transportation.

- Saltation

_____

_____

_____

- Suspension

_____

_____

_____

- Surface creep

_____

_____

_____

4 Explain the interaction between the process of vegetation growth and aeolian processes and how this results in dune accretion.

_____

_____

_____

_____

_____

_____

_____

_____

_____

_____

_____

_____

_____

_____

_____

_____

_____

_____

_____

    PHOTOCOPYING OF THIS PAGE IS RESTRICTED UNDER LAW.   ISBN: 9780170446907

5 How does human use/abuse of frontal dune systems lead to the formation of erosional features such as blowouts and parabolic dunes?

_____

_____

_____

_____

_____

_____

_____

_____

_____

_____

_____

_____

_____

6 In the table below, fully discuss the similarities and differences between marine processes and aeolian processes that operate within a coastal environment.

SIMILARITIES	DIFFERENCES

Introduction to Muriwai Coastal Environment (MCE)

One of New Zealand's most iconic coastlines, Muriwai Beach is located 42 km west from Auckland City. It is a long and very straight coastline, stretching 50 km from Otakamiro Headland in the south to the mouth of the Kaipara Harbour in the north. Highly exposed to the prevailing southwesterly and westerly wind and weather, Muriwai is a rugged high-energy coastline. Waves are large and for this reason it is a popular surf beach. The beach is wide and flat with large foredunes. The southern end of the coastline is made up of tall, rocky, cliffed headlands and bays.

This particular case study will focus on the southernmost stretch of this coastline (see outline map at right). Close analysis will look at a 3 km stretch from Okiritoto Stream in the north to the large headland in the south called Otakamiro Point and Maukatia Bay (also commonly known as Maori Bay). As geographers, we refer to this case study area as the Muriwai Coastal Environment (MCE). This study area has been shaped by numerous interacting coastal processes and hosts a wide array of unique coastal features. The following pages include a map showing the distribution of some of these key features and a range of annotated photographs detailing these features.

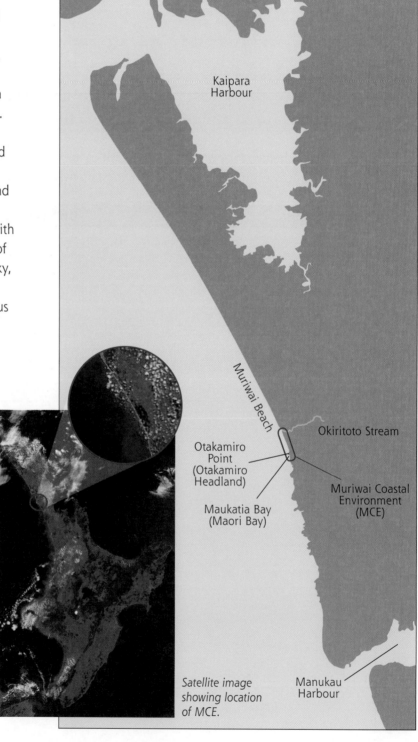

Satellite image showing location of MCE.

PHOTOCOPYING OF THIS PAGE IS RESTRICTED UNDER LAW. ISBN: 9780170446907

Map of Muriwai Coastal Environment showing distribution of selected natural features

Muriwai Coastal Environment's key features

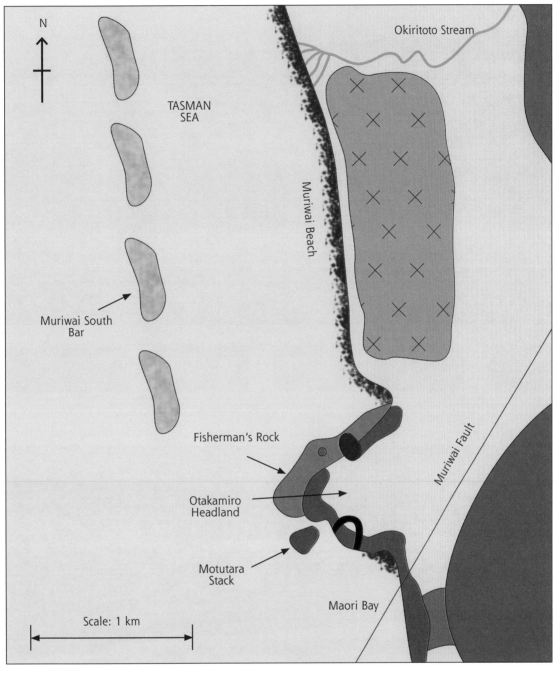

Legend:
- Sedimentary rock platform
- Stack
- Caves
- Blowhole
- Exposed pillow lava
- Ancient forested back dunes
- Old shoreline terrace
- Offshore sand bar
- Sand dunes
- Sea arch
- Iron-sand beach
- Cliffs
- Sandy delta
- River

Aerial view of MCE showing key features

Fisherman's Rock

Otakamiro Headland

Motutara Stack

Maukatia Bay (Maori Bay)

Ancient volcanic cliffs and pillow lava

Muriwai Beach foredunes

Muriwai Beach

    PHOTOCOPYING OF THIS PAGE IS RESTRICTED UNDER LAW.   ISBN: 9780170446907

Tasks

1   In the space below, draw a detailed map of the Muriwai Coastal Environment showing all of its key features. Remember to include all appropriate map conventions (title, scale, north arrow, colour or symbols, key).

Title:

Key:

Key feature 1 – Muriwai Beach

Within our case study environment, the MCE, Muriwai Beach runs for about 2 km from Okiritoto Stream in the north to Otakamiro Headland in the south. The primary process responsible for Muriwai Beach is wave deposition, although in recent years the beach has been heavily eroded.

The beach is very flat (0–5°) and wide (50–100m) – it is referred to as modally dissipative, which means it is a flat winter-style beach for most of the year. This is a direct result of the high levels of wave erosion on the beach due to the large powerful waves.

The sediment primarily comprises very fine black iron-rich sand (less than 0.0025 mm in diameter) known as Mitiwai sediment. The iron component of it is called titanomagnetite. This sand has been eroded by running water from the andesitic central volcanoes of the North Island, as well as Mount Taranaki. Rivers have transported this sediment to the western coastline of New Zealand where longshore drift and littoral drift have shifted this sediment northward to Muriwai Beach and the other west coast beaches. At the MCE, approximately 95 percent of the sediment has originated from this longshore movement of eroded sediment. Only 3 percent arrives via Okiritoto Stream and 2 percent eroded off Otakamiro Headland.

Rates of wave erosion and wave deposition at Muriwai Beach vary temporally between winter and summer. This causes the beach to be very flat in winter and slightly steeper in summer. This will be discussed further in the chapter on temporal variation.

Muriwai Coastal Environment's key features

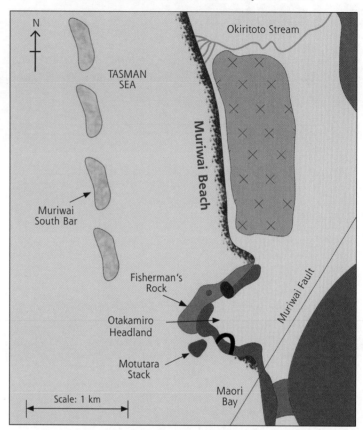

North Island showing origin of Mitiwai sediment at MCE

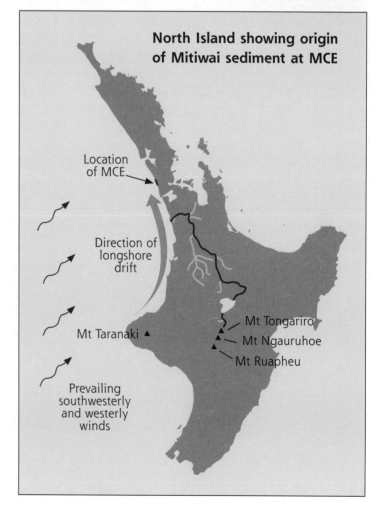

 PHOTOCOPYING OF THIS PAGE IS RESTRICTED UNDER LAW. ISBN: 9780170446907

Key feature 2 – dunes

Muriwai's foredunes run along the whole beach, at the top of the backshore zone. They have largely been formed by aeolian processes associated with strong prevailing onshore winds. They range in height between 5 m and 12 m.

Dunes at Muriwai are heavily vegetated by dune grasses, including spinifex, marram and pikao (pingao). These grasses play an important role in stabilising the dunes, preventing them from being eroded and encouraging them to grow. Most of this vegetation has been planted by local authorities in an effort to maintain a healthy dune system — these dunes act as a natural barrier against the effects of coastal erosion and protect human assets such as houses, car parks and the golf course.

Muriwai Coastal Environment's key features

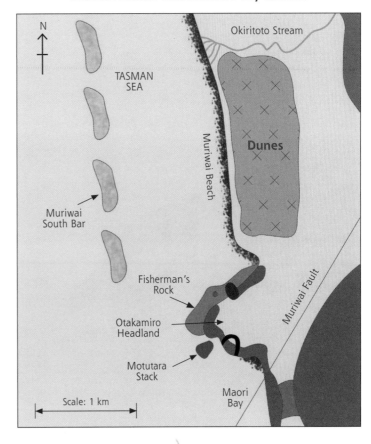

N

TASMAN SEA

Okiritoto Stream

Muriwai Beach

Dunes

Muriwai South Bar

Fisherman's Rock

Otakamiro Headland

Motutara Stack

Muriwai Fault

Maori Bay

Scale: 1 km

Dunes at Muriwai 1.5 km north of Otakamiro Headland

Covered in spinifex grass – this helps to protect the dunes from erosion and encourages deposition of more sand.

Approximately 5 m in height.

In this northern portion of the MCE, the dunes are much more gently sloped than closer to Otakamiro Headland.

Dunes at Muriwai at southern end of beach near Otakamiro Headland

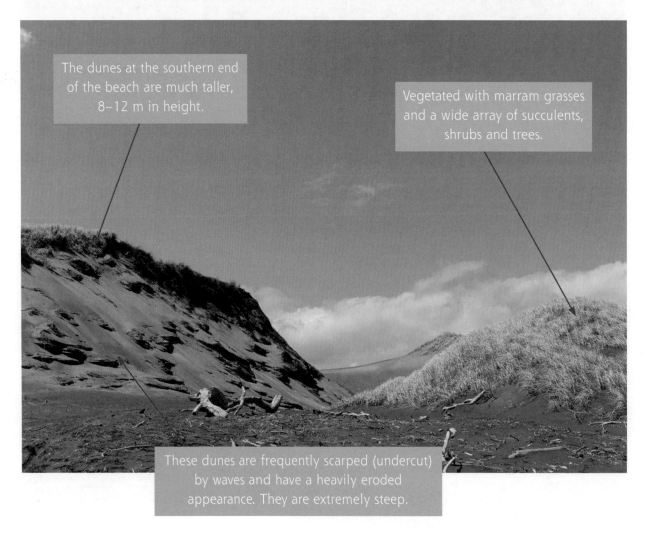

The dunes at the southern end of the beach are much taller, 8–12 m in height.

Vegetated with marram grasses and a wide array of succulents, shrubs and trees.

These dunes are frequently scarped (undercut) by waves and have a heavily eroded appearance. They are extremely steep.

 PHOTOCOPYING OF THIS PAGE IS RESTRICTED UNDER LAW. ISBN: 9780170446907

Different types of dune vegetation at Muriwai Beach

Pikao (pingao).

Marram and spinifex.

Key feature 3 – Otakamiro Headland and its sub-features

Otakamiro Headland is a large cliffed headland protruding from the Muriwai coastline into the Tasman Sea. Its cliffs are approximately 30 m high.

Volcanic activity 17 million years ago was responsible for the hard volcanic conglomerate or Manukau breccia of which Otakamiro Headland is composed. These eruptions came from the Manukau Super Volcano, which laid down much of the geology of the present-day Waitakere Ranges. Much of this activity actually happened under the sea. Tectonic action (predominantly faulting) lifted the headland above sea level 10,000 years ago along the Muriwai–Helensville fault line. This exposed the headland to the action of marine erosion. However, due to the resistant nature of the rock, the headland has eroded very slowly and formed features such as caves.

Otakamiro Headland also has a large flat shore platform known as Fisherman's Rock on its northern side. This shore platform is the result of 10,000 years of marine erosion and sub-aerial weathering. These interacting processes have also created a highly pronounced wave-cut notch, three caves (two of which now join together, beginning the formation of an arch) and the iconic stack formation known as Motutara Stack.

Muriwai Coastal Environment's key features

ISBN: 9780170446907   PHOTOCOPYING OF THIS PAGE IS RESTRICTED UNDER LAW.

Otakamiro Headland at MCE

Otakamiro
Headland

Cliffs, 10–15 m
in height

Large cave on
northern side of the
headland

Fisherman's Rock
(shore platform)

Otakamiro Headland and Motutara Stack

Vegetation – flax and grass

Gannet colony

Motutara
Stack

Otakamiro
Headland

Notch

Maori Bay

PHOTOCOPYING OF THIS PAGE IS RESTRICTED UNDER LAW.   ISBN: 9780170446907

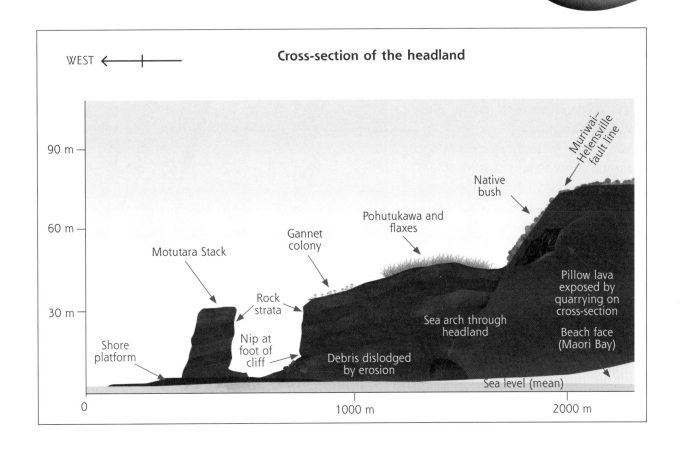

Cross-section of the headland

WEST ←

90 m

60 m

30 m

Muriwai–Helensville fault line

Native bush

Pohutukawa and flaxes

Gannet colony

Motutara Stack

Rock strata

Pillow lava exposed by quarrying on cross-section

Sea arch through headland

Beach face (Maori Bay)

Shore platform

Nip at foot of cliff

Debris dislodged by erosion

Sea level (mean)

0   1000 m   2000 m

Aerial view of Fisherman's Rock at Otakamiro Headland

Motutara Stack

Platform

Otakamiro Headland

Gannet colony

Notch

Key feature 4 – Maukatia Bay (Maori Bay)

South of Otakamiro Headland is a continuous chain of bays and headlands running along the coastal margin of the Waitakere Ranges.

Within our MCE case study area is Maukatia Bay (also known as Maori Bay). This is a cliffed small bay, surrounded by the same rock found at Otakamiro Headland — produced by the Manukau Super Volcano 17 million years ago. Because of the tall steep cliffs, sand dunes have not formed here in the same way they have on Muriwai Beach.

Much of the cliffs were formed underwater. As the lava extruded slowly from volcanic vents and cooled rapidly in the water, distinct geological formations known as pillow lava were formed.

Muriwai Coastal Environment's key features

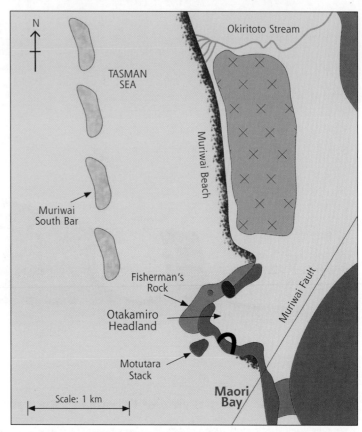

Maori Bay (looking south from Otakamiro Headland)

PHOTOCOPYING OF THIS PAGE IS RESTRICTED UNDER LAW.   ISBN: 9780170446907

Pillow lava formations on a cliff at Maukatia Bay (Maori Bay).

KEY ELEMENTS OF MCE

(Elements are the key building blocks of the environment that make up the features and contribute to how processes work)

Waves

Average wave height is 1.5 m with an average energy of 250 000 joules per m³. Large swells generated in the Southern Ocean near Antarctica travel a lengthy fetch of up to several thousand kilometres, spurred on by constant prevailing southwesterly and westerly winds. Waves are often large, foamy, spilling waves, however often at times can be plunging and highly erosive.

The size, energy and erosivity of waves vary significantly between summer and winter.

Geology

At the southern end of the MCE at Otakamiro Point (Headland) and Maukatia Bay (Maori Bay), the rock is of igneous origin, dating back 17 million years. Most of the rock here is a conglomerate known as Manukau Breccia, comprising stone such as andesite and basalt. It is extremely erosion resistant (low erodibility).

North of Otakamiro Point, much softer sandstone and siltstone has been eroded back prior to the deposition of Muriwai Beach and formation of an extensive dune system (which occurred approximately 2 million years before present).

Vegetation

Early missionaries to the area in the 1800s reported that the dunes were mostly unvegetated, several kilometres broad and shifting constantly. Probably the dunes were covered in forest before the first Maori arrived.

Nowadays the dunes are extensively vegetated in native and introduced grasses, such as spinifex and marram. Further back on the ancient dunes, humans have established a golf course and the plantation forest known as Woodhill. Cliffs are vegetated with grasses, flax and trees such as pohutukawa.

Wind

The prevailing wind direction is southwesterly, with westerly being the second most predominant wind direction. Southwesterly and westerly winds blow 'onshore' – this is important in terms of the formation and shaping of dunes.

The average wind velocity is 6 m per second. The fluid threshold velocity for Mitiwai sediment (speed required for sediment erosion) is 5.47 m per second and winds blow over this threshold 67 percent of the time.

(continued next page)

Sediment

The fine black sand of Muriwai is known as Mitiwai sediment. It is iron rich, comprising titanomagnetite. Average particle size is 0.0025 mm. This sediment originates from erosion of central and west coast volcanoes of the North Island and is transported by longshore processes up the west coast in large 'slugs' of up to 350 000 m³; 95 percent of the sediment at MCE comes from the volcanoes and longshore transport, 3 percent is discharged from Okiritoto Stream and 2 percent originates from Otakamiro Headland.

Humans

Humans play a significant role in the environment. There is a large township located at the southern end of Muriwai Beach near Otakamiro Headland and the ancient dune systems are partly covered by a golf course and a human-planted forest called Woodhill. The beach is extensively used for recreation, and vehicles are permitted. Humans have intentionally and unintentionally had a huge impact on how natural processes operate and interact. In recent years, humans have tried hard to manage and mitigate erosion of the coast.

Tasks

The following questions refer to important specific detail you need to know well about the different features at Muriwai Coastal Environment. Read pages 40–52 to find out this key information.

1 Approximately how long is the stretch of Muriwai Beach between Otakamiro Headland and Okiritoto Stream?

2 How wide is Muriwai Beach and what is the slope gradient?

3 Muriwai Beach is referred to as being 'modally dissipative'. What does this mean?

4 What is the average grain size of Mitiwai sand at Muriwai Beach?

5 What is the iron component of Mitiwai sand known as?

6 What colour is Mitiwai sand?

7 Name four North Island volcanoes where Mitiwai sand originates from.

8 Name two processes that have transported Mitiwai sediment along the North Island's west coast towards the MCE.

9 Name two other sources of sediment at MCE.

PHOTOCOPYING OF THIS PAGE IS RESTRICTED UNDER LAW.   ISBN: 9780170446907

10 How high are Muriwai Beach's frontal dunes?

_____

11 Name three specific types of dune vegetation that grow on the frontal dunes at Muriwai Beach.

_____

12 Why are the frontal dunes at Muriwai Beach considered important by humans?

_____

13 Approximately how high is Otakamiro Headland?

_____

14 What natural process formed the initial rock substrate of Otakamiro Headland (under the Tasman Sea)?

_____

15 What natural process uplifted Otakamiro Headland above the surface of the sea?

_____

16 Why does Otakamiro Headland erode much slower than other parts of the MCE's coastline?

_____

17 What is the name of the volcanic rock that makes up most of Otakamiro Headland?

_____

18 What is the name of the shore platform at Otakamiro Headland?

_____

19 What is the name of the stack at Otakamiro Headland?

_____

20 What is another name for Maukatia Bay?

_____

21 What fringes the backshore of Maukatia Bay instead of sand dunes?

_____

22 What is the name of the unique lava formations found at Maukatia Bay?

_____

4

How have interacting natural processes shaped the key features at Muriwai Coastal Environment (MCE)?

Muriwai Beach

Muriwai Beach is a depositional feature. Even though the beach goes through intense periods of erosion, it was the process of wave deposition that put it there in the first place.

Geographers are not certain as to when Muriwai Beach first formed, although it is somewhere in the timeframe of tens of thousands of years. As already stated, 95 percent of the sediment is brought into the environment by the

process of littoral drift. Large 'slugs' of sediment take thousands of years to migrate from the south up into the MCE. When the sediment drifts past Otakamiro Headland, it interacts with the waves and is deposited on the coast. Wave deposition builds up the beach.

Contrary to this, the beach is also shaped by wave erosion, which can erode the beach backwards and flatten it out. Longshore drift also plays a role in distributing sediment along Muriwai Beach. These phenomena will be discussed further in the next two chapters.

Key message
Muriwai Beach is a depositional feature. Initially formed by the interaction of littoral drift and wave deposition, it is also shaped by wave erosion and longshore drift.

PHOTOCOPYING OF THIS PAGE IS RESTRICTED UNDER LAW.   ISBN: 9780170446907

Muriwai Beach foredunes

Dunes are shaped at Muriwai Coastal Environment (MCE), along the backshore of Muriwai Beach, by the interaction of marine processes, aeolian processes and vegetation processes.

Littoral drift supplies the MCE with titanomagnetite sands, which are then deposited onto Muriwai Beach by wave deposition. At the foreshore, sediment is then eroded by wind — a process known as deflation. For high rates of deflation to occur, the following conditions must be present:

- Low tide, so large amounts of sediment are exposed to the wind.
- Sediment is friable (loose, unconsolidated) — the sediment must be dried out by the sun, as wet sediment is too heavy and sticky.
- Winds blowing above the fluid threshold velocity for MCE — this is 5.47 metres per second, or roughly 20 kilometres per hour. Winds blow over this threshold 67 percent of the time.
- Wind must be blowing onshore if sediment is to be transported towards the dunes. Winds blow this direction (southwesterly and westerly) 59 percent of the time at Muriwai Beach.

Once entrained, sediment travels up the beach slope through the processes of saltation, suspension and surface creep. Vegetation growth occurs on the dunes at Muriwai and this process interacts with aeolian transportation. Grasses growing there, such as spinifex, marram and pikao, slow down wind speeds below the 5.47 metres per second, causing aeolian deposition to occur.

> **Key message**
> Dunes are formed and shaped through the interaction of marine, aeolian and vegetation processes. At Muriwai Beach, these interacting natural processes have caused the dunes to increase in size.

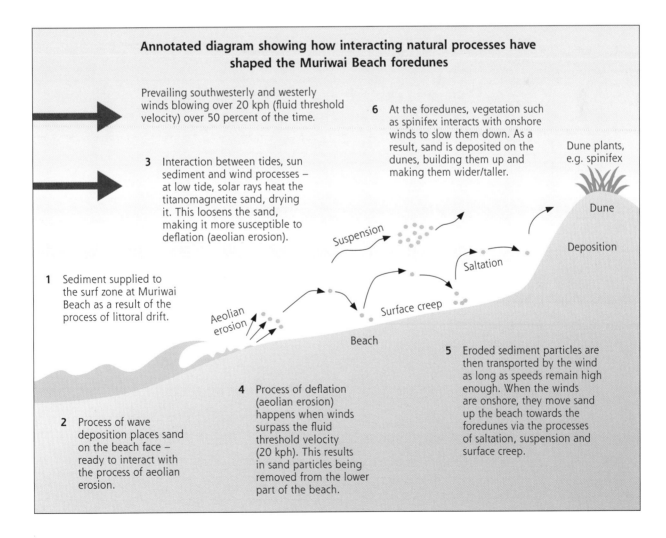

Annotated diagram showing how interacting natural processes have shaped the Muriwai Beach foredunes

Prevailing southwesterly and westerly winds blowing over 20 kph (fluid threshold velocity) over 50 percent of the time.

3 Interaction between tides, sun sediment and wind processes – at low tide, solar rays heat the titanomagnetite sand, drying it. This loosens the sand, making it more susceptible to deflation (aeolian erosion).

6 At the foredunes, vegetation such as spinifex interacts with onshore winds to slow them down. As a result, sand is deposited on the dunes, building them up and making them wider/taller.

Dune plants, e.g. spinifex

Dune

Deposition

Suspension

Saltation

1 Sediment supplied to the surf zone at Muriwai Beach as a result of the process of littoral drift.

Aeolian erosion

Surface creep

Beach

5 Eroded sediment particles are then transported by the wind as long as speeds remain high enough. When the winds are onshore, they move sand up the beach towards the foredunes via the processes of saltation, suspension and surface creep.

2 Process of wave deposition places sand on the beach face – ready to interact with the process of aeolian erosion.

4 Process of deflation (aeolian erosion) happens when winds surpass the fluid threshold velocity (20 kph). This results in sand particles being removed from the lower part of the beach.

Tasks

1 Discuss in several sentences how the processes of littoral transportation and wave deposition have interacted to form Muriwai Beach.

_____

_____

_____

_____

_____

_____

_____

_____

_____

_____

_____

2 Explain why deflation and aeolian transportation occur at high rates at Muriwai Beach. Include case study specific detail in your answer.

_____

_____

_____

_____

_____

_____

_____

_____

_____

_____

_____

_____

_____

_____

_____

_____

_____

_____

_____

 PHOTOCOPYING OF THIS PAGE IS RESTRICTED UNDER LAW. ISBN: 9780170446907

3 Draw a detailed annotated diagram that illustrates how interacting natural processes shape the frontal dunes *at Muriwai Beach. Make sure you actually discuss these interactions in your annotations.*

Otakamiro Headland and its many features

Otakamiro Headland has been formed over a long period of millions of years. A number of different processes have interacted to form and shape it. This began with undersea volcanism to lay down the rock, and then a combination of tectonic uplift (faulting along the Muriwai–Helensville fault line) and sea-level change to expose the headland above the Tasman Sea. Over tens of thousands of years, the headland has been slowly modified by the interaction of marine erosion and sub-aerial weathering to produce its distinct features we see today, including its tall cliffs, three caves, and extensive shore platform known as Fisherman's Rock and Motutara Stack. The following series of diagrams detail the changes that have taken place over time, and what could happen in the future.

Diagrams showing the shaping of Otakamiro Headland from 20 million years before present to nowadays

20 Million Years Before Present (YBP)

An undersea volcano laid down the Manukau Breccia and Piha conglomerates that form the substructure of Otakamiro Headland.

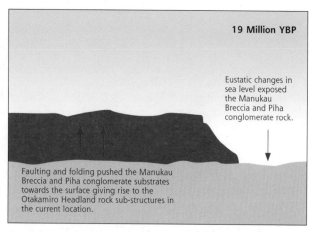

19 Million YBP

Eustatic changes in sea level exposed the Manukau Breccia and Piha conglomerate rock.

Faulting and folding pushed the Manukau Breccia and Piha conglomerate substrates towards the surface giving rise to the Otakamiro Headland rock sub-structures in the current location.

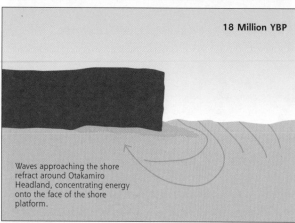

18 Million YBP

Waves approaching the shore refract around Otakamiro Headland, concentrating energy onto the face of the shore platform.

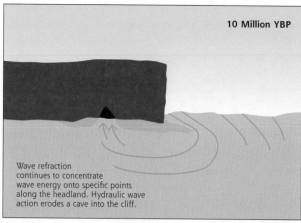

10 Million YBP

Wave refraction continues to concentrate wave energy onto specific points along the headland. Hydraulic wave action erodes a cave into the cliff.

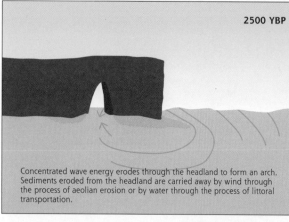

2500 YBP

Concentrated wave energy erodes through the headland to form an arch. Sediments eroded from the headland are carried away by wind through the process of aeolian erosion or by water through the process of littoral transportation.

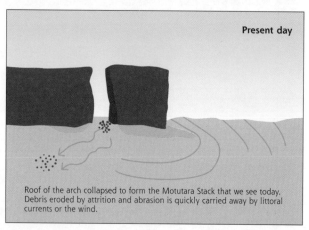

Present day

Roof of the arch collapsed to form the Motutara Stack that we see today. Debris eroded by attrition and abrasion is quickly carried away by littoral currents or the wind.

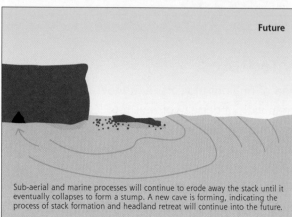

Future

Sub-aerial and marine processes will continue to erode away the stack until it eventually collapses to form a stump. A new cave is forming, indicating the process of stack formation and headland retreat will continue into the future.

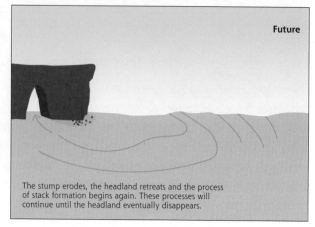

Future

The stump erodes, the headland retreats and the process of stack formation begins again. These processes will continue until the headland eventually disappears.

 PHOTOCOPYING OF THIS PAGE IS RESTRICTED UNDER LAW. ISBN: 9780170446907

Fisherman's Rock has been formed over this same time period by concentrated wave erosion (hydraulic action and abrasion) occurring at the foot of the cliff, at the mean high tide mark. This repetitive process of erosion ate away at a notch at the base of the cliff causing the cliff to become unstable and collapse. As this process has continued over the last 2 million years, Fisherman's Rock has formed.

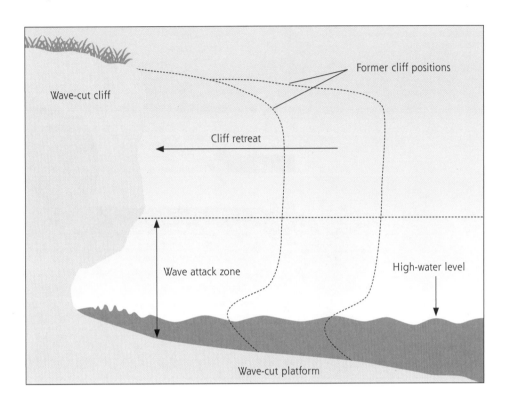

Sub-aerial weathering also contributes to shaping the headland, its cliffs and the platform. Salt crystallisation on the headland is prolific. This weathering is important because it weakens the rock, increasing rates of marine erosion — a classic example of processes interacting.

One key process of sub-aerial weathering that occurs on Otakamiro Headland is known as 'spray-splash weathering'. This form of erosion occurs above the high-water mark, high up on the cliffs by spray and splash produced by breaking waves. The cliffs are wetted with ocean spray, which in turn dry out leaving behind salt crystals that eat away at the rock. This produces the distinctive honeycomb pattern typical of spray-splash weathering. It is generally not as important as the hydraulic action by waves in causing the retreat of cliffs behind shore platforms like Fisherman's Rock, yet it speeds up the process.

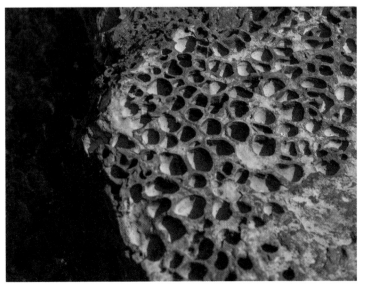

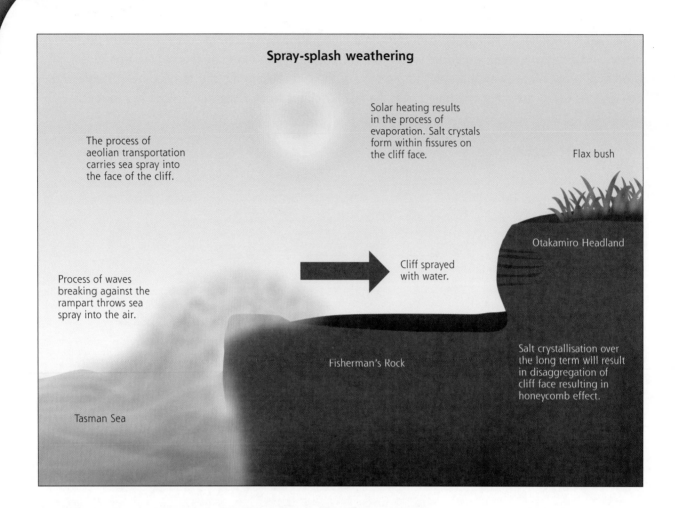

Spray-splash weathering

The process of aeolian transportation carries sea spray into the face of the cliff.

Solar heating results in the process of evaporation. Salt crystals form within fissures on the cliff face.

Flax bush

Otakamiro Headland

Cliff sprayed with water.

Process of waves breaking against the rampart throws sea spray into the air.

Fisherman's Rock

Salt crystallisation over the long term will result in disaggregation of cliff face resulting in honeycomb effect.

Tasman Sea

Another important form of sub-aerial weathering that shapes Otakamiro Headland is 'water-layer weathering'. The wetting and drying of Fisherman's Rock's surface due to tidal changes and different sized waves can cause this type of weathering. As the tide retreats, it leaves salt-water ponds behind, which the sun's heat evaporates, leaving behind salt in the form of crystals. As these crystals grow they flake off rock, gradually eroding the platform surface (lowering it).

A diagram showing this process can be seen on the next page.

PHOTOCOPYING OF THIS PAGE IS RESTRICTED UNDER LAW.
ISBN: 9780170446907

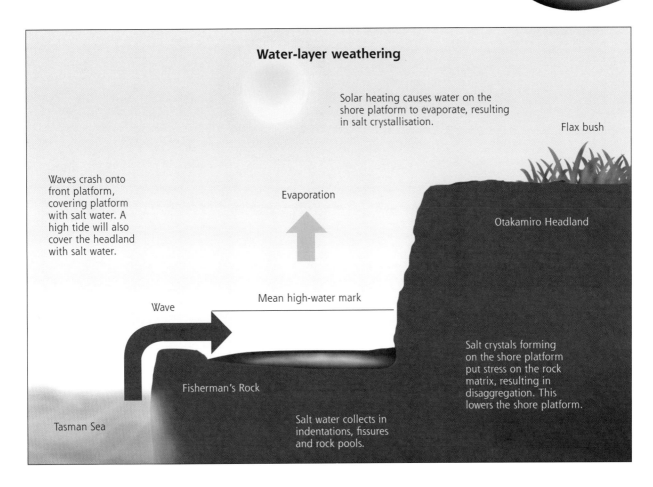

Water-layer weathering

Solar heating causes water on the shore platform to evaporate, resulting in salt crystallisation.

Flax bush

Evaporation

Waves crash onto front platform, covering platform with salt water. A high tide will also cover the headland with salt water.

Otakamiro Headland

Mean high-water mark

Wave

Fisherman's Rock

Salt crystals forming on the shore platform put stress on the rock matrix, resulting in disaggregation. This lowers the shore platform.

Tasman Sea

Salt water collects in indentations, fissures and rock pools.

 Tasks

1 Construct a series of *three* detailed annotated diagrams that show how interacting natural processes have interacted to shape Motutara Stack at Otakamiro Headland.

Stage 1:

Stage 2:

Stage 3:

INTERACTING NATURAL PROCESSES IN A COASTAL ENVIRONMENT   PHOTOCOPYING OF THIS PAGE IS RESTRICTED UNDER LAW.   ISBN: 9780170446907

2 Construct a detailed annotated diagram showing how spray-splash weathering and water-layer weathering have shaped Fisherman's Rock.

ISBN: 9780170446907   PHOTOCOPYING OF THIS PAGE IS RESTRICTED UNDER LAW.

5 Temporal variation of interacting natural processes at Muriwai Coastal Environment (MCE)

Temporal variation = how something varies (changes) over time or between different time periods.

In the Muriwai Coastal Environment (MCE), there are many different instances of where the dominant coastal processes vary temporally. This chapter will closely analyse *three* of these variations and how they subsequently shape the natural environment and its features:

1 Seasonal variation of wave processes and how this shapes the beach profile.
2 Temporal variation of aeolian processes between 1950 and present.
3 Longer term temporal variation of alternating erosional and depositional periods affecting the MCE over thousands of years.

Temporal variation 1 — seasonal variation of wave processes

Winter

During winter months (July, August), winds blow at a much higher velocity due to the higher occurrence of low air-pressure systems. Large storms are very common, both in the Tasman Sea and far off in the Southern Ocean where Muriwai's waves are generated. The climatic processes of winter, including wind and storm processes, interact with wave processes, causing waves to be bigger, steeper and more destructive.

Winter waves would typically have over *400 000 joules per m³* of energy with heights over *3 m*. These waves *plunge* onto the beach and a higher ratio of backwash to swash results in daily beach erosion. Mitiwai sand stripped from the beach is usually transported via suspension in *rip currents*, where it is deposited in the offshore zone forming *transverse bars*.

Key interacting processes:
Climatic processes (wind processes) *interact with* wave processes (wave erosion and wave deposition).

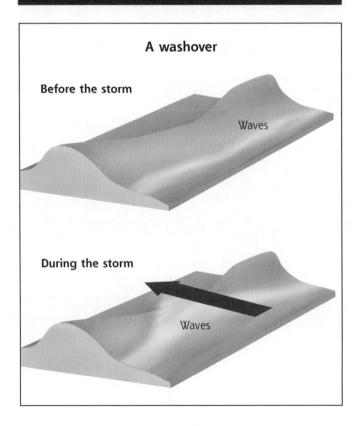

A washover

Before the storm

Waves

During the storm

Waves

 PHOTOCOPYING OF THIS PAGE IS RESTRICTED UNDER LAW. ISBN: 9780170446907

These higher rates of erosion in the winter have a distinct impact on the shape of Muriwai Beach and affect the beach profile (what the beach looks like from side on). In the middle of winter, Muriwai Beach is its *flattest* throughout the year, with a gradient of 0–2.5°. This is because the top layers of sediment, including the berm, have been eroded and transported away.

This incredibly flat beach also allows for wave erosion of the frontal dunes during very high tides and storms. The dunes can become undercut (scarped). Washovers can occur where waves literally wash over the dune at various points, eroding out wide channels.

Summer

The interaction between climatic and marine processes varies significantly in summer from how they interact in winter and this has a distinct shaping effect on the environment.

In summer (January, February), waves hitting Muriwai Beach can be significantly more gentle when compared with winter waves. This is due to how waves interact with calmer wind conditions in these months. The more gentle winds are associated with the high-pressure atmospheric conditions that the southern hemisphere experiences in summer.

These gentler waves tend to possess energy between *25 000 and 100 000 joules per m³* of energy with heights *under 1.5 m*. They tend to be *spilling*, as opposed to plunging. They swash up the beach. Backwash is minimised by much of the water infiltrating (soaking) into the beach. Waves carry sediment in them, held in suspension, and as they swash up the beach they *deposit material*. Much of this material has originated from the offshore zone, reducing the size of offshore bars.

In the middle of summer, Muriwai Beach is significantly *steeper* when compared to winter, with a gradient of 5–10°. Sediment has been deposited on the beach, building it up and making it steeper. A berm is evident in the summer. Also during summer months, dunes are less likely to be eroded by waves. In fact, dunes can experience significant growth in these months — drier weather means sediment is more friable and therefore much more easily eroded and saltated by wind. Dunes stabilise during summer.

Below are diagrams of beach profiles showing the difference between winter and summer.

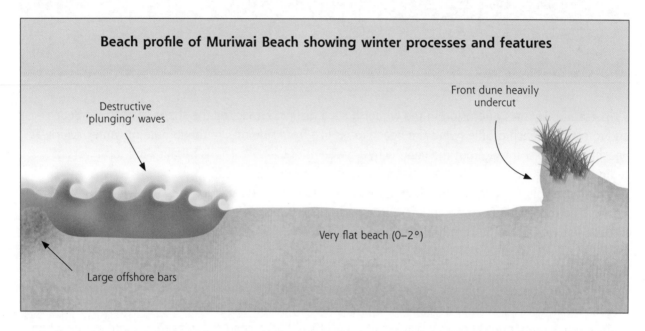

Beach profile of Muriwai Beach showing winter processes and features

Destructive 'plunging' waves

Front dune heavily undercut

Very flat beach (0–2°)

Large offshore bars

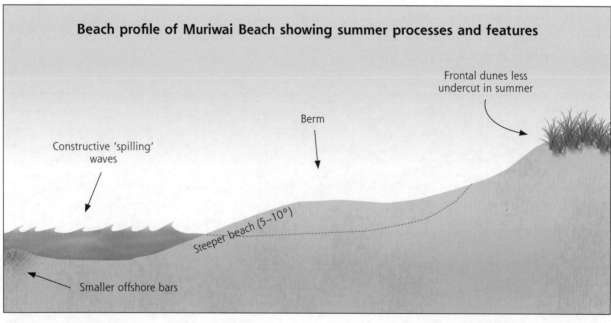

Beach profile of Muriwai Beach showing summer processes and features

Frontal dunes less undercut in summer

Berm

Constructive 'spilling' waves

Steeper beach (5–10°)

Smaller offshore bars

 PHOTOCOPYING OF THIS PAGE IS RESTRICTED UNDER LAW. ISBN: 9780170446907

Tasks

1   Describe and explain how wind differs between summer *and* winter at MCE. Use the term 'climatic processes' in your answer.

_____

_____

_____

_____

_____

_____

_____

_____

_____

_____

_____

_____

_____

_____

_____

2   Describe marine processes at MCE in summer, compared to winter. How does this show interaction with climatic processes?

_____

_____

_____

_____

_____

_____

_____

_____

_____

_____

_____

_____

_____

_____

_____

_____

_____

_____

_____

_____

3 Fully explain how this seasonal variation in climatic processes interacts with marine processes to shape the MCE, in particular the beach and the dunes. Include specific case study detail.

BEACH:

_____
_____
_____
_____
_____
_____
_____
_____
_____
_____
_____
_____
_____
_____
_____
_____
_____
_____
_____

DUNES:

_____
_____
_____
_____
_____
_____
_____
_____
_____
_____
_____
_____
_____
_____
_____
_____
_____
_____
_____
_____

    PHOTOCOPYING OF THIS PAGE IS RESTRICTED UNDER LAW.   ISBN: 9780170446907

4 Draw an annotated diagram (or diagrams) to fully explain seasonal variation of interacting natural processes and how this shapes the environment at the MCE. Make sure to include highly detailed annotations.

Temporal variation 2 — variation in aeolian processes between 1950 and present

Key interacting processes:
Processes of vegetation growth *interact with* aeolian deposition and aeolian erosion.

As we already know, dunes are shaped at Muriwai as the result of a sequence of interacting processes:

1 At low tide, solar heating dries out sediment on the beach making it friable.
2 Onshore winds with a velocity over the fluid threshold velocity (5.47 mps) are able to deflate (erode) sediment from the foreshore.
3 Sediment saltates (bounces) up the beach face towards the backshore zone.
4 In the backshore zone, at the frontal dunes, vegetation growth (e.g. spinifex and marram) slows down wind flow resulting in aeolian deposition.
5 This deposition results in dune accretion, making the dunes taller and wider.

However, this generalised model of how dunes are shaped varied in a significant way between the 1950s and present at the MCE, largely due to human modification of the process of vegetation growth.

How were the processes operating in 1950?

Between 1840 and 1950, the removal of dune vegetation resulted in loss of frontal dunes due to reduced deposition and increased erosion.

Seven to eight hundred years ago Muriwai was heavily settled by Maori. These first settlers extensively removed coastal vegetation (mainly by burning). The first British settlers cut away even more vegetation in order to establish new farmland. In the 1840s, the Reverend Samuel Marsden described the dunes at Muriwai Beach as follows: '... the sandhills are very high ... There is no vegetation on them and the sand shifts with the contending winds ...'

By the 1920s, farmers were grazing cattle on the Muriwai dunes, further reducing the vegetation cover. The lack of vegetation on the dunes meant two things:

- Rates of aeolian deposition at the dunes dropped dramatically due to the lack of interaction with vegetation. This meant existing dunes were not growing, and new dunes were not forming.
- Aeolian erosion at the dunes actually increased to higher rates than deposition. This meant existing dunes at the beach were being swept away and transported far inland. The beach lost its dunes. In 1880, the area of drifting sands was estimated at 40 000 ha and had increased to 120 000 ha by 1909.

By 1950, the dunes were non-existent. Sediment was being blown up to 5 km inland. A lack of frontal dunes accelerated marine erosion as the beach was without its natural defences. Many human assets, including the golf course and numerous buildings, were under significant threat from the action of the waves.

 PHOTOCOPYING OF THIS PAGE IS RESTRICTED UNDER LAW. ISBN: 9780170446907

How do the interacting natural processes vary today compared with the 1950s?

The operation of aeolian processes today varies significantly from the early 1900s. Extensive coastal management and dune stabilisation has been undertaken by the Auckland Council since 1950. Marram and then spinifex was planted on the frontal dunes. This vegetation growth has interacted with aeolian processes, allowing for less erosion of the dunes as the roots of the grasses bind the sediment together and act to slow down wind speed. Vegetation growth has also increased rates of aeolian deposition — resulting in dune accretion (build-up) as wind-eroded sediment is dropped and left to settle around the grasses. Muriwai Beach nowadays has extensive, healthy dune systems, which help to protect the coastline from increased erosion. Dunes range in height from 3 m to 8 m and provide much more adequate coastal defence from waves when compared with the 1950s. The following diagrams showing the variation.

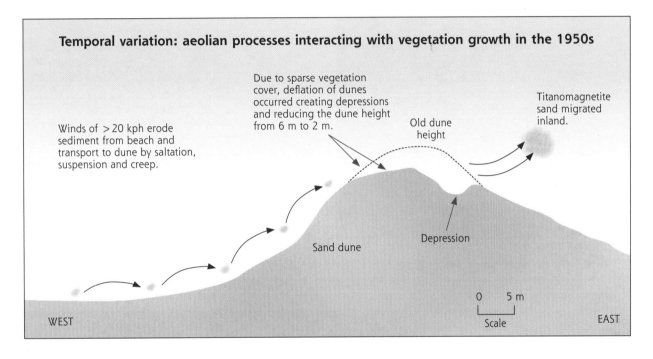

Temporal variation: aeolian processes interacting with vegetation growth in the 1950s

Due to sparse vegetation cover, deflation of dunes occurred creating depressions and reducing the dune height from 6 m to 2 m.

Winds of > 20 kph erode sediment from beach and transport to dune by saltation, suspension and creep.

Titanomagnetite sand migrated inland.

Old dune height

Sand dune

Depression

WEST

0   5 m
Scale

EAST

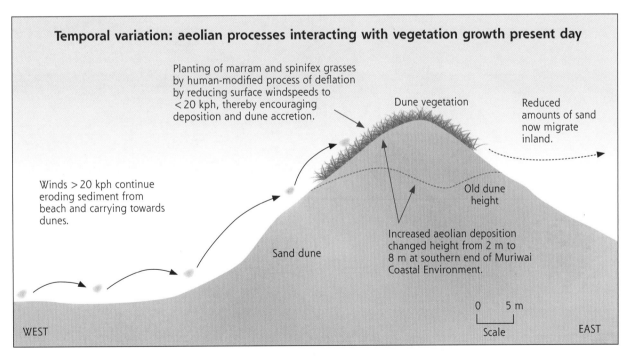

Temporal variation: aeolian processes interacting with vegetation growth present day

Planting of marram and spinifex grasses by human-modified process of deflation by reducing surface windspeeds to < 20 kph, thereby encouraging deposition and dune accretion.

Dune vegetation

Reduced amounts of sand now migrate inland.

Winds > 20 kph continue eroding sediment from beach and carrying towards dunes.

Old dune height

Sand dune

Increased aeolian deposition changed height from 2 m to 8 m at southern end of Muriwai Coastal Environment.

WEST

0   5 m
Scale

EAST

Images below left (1951) and below right (1957) show a landscape that has been extensively stripped of most vegetation that helps to stabilise the dunes, which in turn led to their erosion.

1951

1957

Today (below), even though we can see the presence of buildings, infrastructure and a golf course, the dunes themselves have been heavily vegetated with plants such as spinifex and marram, which have stabilised the dunes and encouraged deposition of sediment, building up the dunes.

 PHOTOCOPYING OF THIS PAGE IS RESTRICTED UNDER LAW. ISBN: 9780170446907

Tasks

1 Complete the table to fully explain how interacting processes of vegetation growth and aeolian erosion/deposition have varied at the MCE between 1950 and the present.

	1950	Present day
Explain how the process of vegetation growth was operating at MCE dunes.		
How was this a direct result of human management/ mismanagement?		
How did the process of vegetation growth during this period interact with aeolian processes?		
How did this shape the environment, in particular the frontal dunes?		

2 Draw an annotated diagram showing the interaction of vegetation processes and the aeolian processes for each time period.

1950

Present day

 PHOTOCOPYING OF THIS PAGE IS RESTRICTED UNDER LAW. ISBN: 9780170446907

Temporal variation 3 — long-term variation of erosional and depositional phases over thousands of years

The Muriwai coastline is an open system where sediment comes and goes. It is rarely in a state of equilibrium. Between 4000 and 2000 years ago Muriwai was growing due to a higher ratio of wave deposition to erosion. However, 2000 years ago to present, Muriwai Beach has been experiencing an erosional phase.

In order to understand why Muriwai experiences these long-term depositional and erosional phases, we must understand where the sediment comes from and how it gets to the MCE. You will remember from Chapter 3 of this book that 95 percent of Muriwai's sediment, its titanomagnetite compounds, comes from the North Island's big volcanoes — Taranaki, Ruapehu, Tongariro and Ngauruhoe (see map, right). Rivers and streams transport material eroded from these volcanoes to the coast, where it is discharged into the sea along the coast. Because of New Zealand's prevailing southwesterly winds, longshore currents and the processes of longshore drift and littoral drift move this material northward towards the MCE, and other west coast beaches such as Piha.

Sediment being transported up the North Island's west coast does not move in a constant, uniform 'conveyor belt' motion. Instead it pulses its way up the coast in giant masses of sediment known as 'slugs'. These slugs can be up to 3500 km^3 in size and take hundreds, if not thousands, of years to move along the coast. The sediment slug's journey up the coast will stop when it reaches specific obstacles or barriers, such as river mouths, harbour mouths or large headlands. Slugs will sit and wait here and collect more and more sediment from the continued process of longshore and littoral drift, until eventually they gather enough mass to be able to pass by such obstacles. As a slug of sediment passes by a particular section of the North Island's western coastline, it can have a profound interaction with wave processes, increasing rates of wave deposition.

> **Key interacting processes:**
> Littoral movement (longshore drift and littoral drift) *interacts with* wave processes (wave erosion and wave deposition).
> Wave deposition *interacts with* aeolian erosion and deposition.

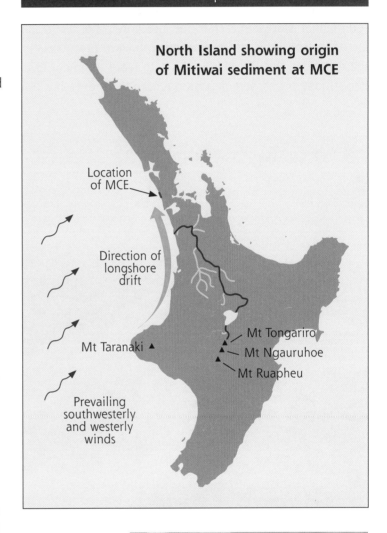

North Island showing origin of Mitiwai sediment at MCE

Location of MCE

Direction of longshore drift

Mt Taranaki ▲

Mt Tongariro
Mt Ngauruhoe
Mt Ruapheu

Prevailing southwesterly and westerly winds

Fine black Mitiwai sediment being deposited at MCE.

Muriwai Beach 4000–2000 years ago: depositional phase

Between 4000 and 2000 years ago, Muriwai Beach experienced a distinct depositional phase. It is estimated that a depositional region 10–15 km wide built its way out from the Muriwai hills. A huge wide beach area was formed, and aeolian processes created an extensive dune system. Today we can still see this area — it is the very large area of ancient dunes covered by the golf course and Woodhill Forest.

This large-scale depositional phase was the result of a higher input of sediment into the environment compared to outputs (sediment leaving the MCE and moving to other coastal environments to the north). This is because a major slug (or series of slugs) had made its way up the west coast and was pushing past Otakamiro Headland. Features such as Okiritoto Stream and the mouth of the Kaipara Harbour were next in line to halt the movement of this slug, and the result was a net increase in sediment at Muriwai Beach in the foreshore zone. Technically, rates of littoral sediment movement were increasing, and this was interacting with wave processes — by increasing supply of sediment going into the foreshore, this provided waves with more sediment to transport to the coastline — increasing wave deposition. Waves were depositing more sediment, thus building up the coastline.

Aerial photograph of Muriwai Beach – note the large area behind the beach containing the golf course, Woodhill Forest and Muriwai township. This area of ancient dunes was built up by wave deposition during the long-term period of deposition 4000–2000 years ago.

Muriwai Beach 2000 years ago to present: erosional phase

Approximately 2000 years ago, the sediment slug that had been feeding the MCE with fresh new sediment began to move on further up the coastline past the Kaipara Harbour. As a result, rates of littoral movement of sediment into the MCE dropped significantly. This interacted with wave processes by supplying waves with less sediment, reducing rates of wave deposition. Waves were able to erode more and longshore drift helped to carry Muriwai's sediment northwards out of the environment.

PHOTOCOPYING OF THIS PAGE IS RESTRICTED UNDER LAW.   ISBN: 9780170446907

All of this resulted in a long-term erosional phase, causing coastal retreat. During the later 1900s and early 2000s, geographers measured that Muriwai Beach was eroding back at a rate of 1.4 m per year. During storms the beach would become significantly narrower and allow for the erosion of foredunes.

Interestingly, since 2015 coastal managers working at Muriwai have identified that the erosional retreat of the beach has slowed significantly, to the point where it is reversing. It is believed that Muriwai is heading into a new long-term depositional phase as a new slug of west coast sediment has moved along the coast and is beginning to break through into the MCE.

Depositional phase
4000–2000 years ago

Erosional phase
2000 years ago to present day

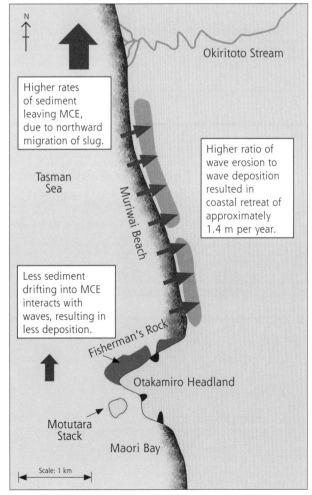

Okiritoto Stream

Lower rates of sediment leaving MCE.

Higher ratio of wave deposition to wave erosion during this period results in beach growing outwards.

Tasman Sea

Muriwai Beach

Higher rates of sediment drifting into MCE due to presence of slug.

Fisherman's Rock

Otakamiro Headland

Motutara Stack

Maori Bay

Scale: 1 km

Okiritoto Stream

Higher rates of sediment leaving MCE, due to northward migration of slug.

Higher ratio of wave erosion to wave deposition resulted in coastal retreat of approximately 1.4 m per year.

Tasman Sea

Muriwai Beach

Less sediment drifting into MCE interacts with waves, resulting in less deposition.

Fisherman's Rock

Otakamiro Headland

Motutara Stack

Maori Bay

Scale: 1 km

◉ Tasks

1 Explain what a 'slug' is and describe its movement northwards along New Zealand's west coast.

_____

_____

_____

_____

_____

_____

_____

_____

2 Explain how this slug shaped Muriwai Beach 4000–2000 years ago.

_____

_____

_____

_____

_____

_____

_____

3 Explain how this slug was shaping Muriwai Beach 2000 years ago to present.

_____

_____

_____

_____

_____

_____

4 How can this variation be explained by looking at littoral transportation interacting with wave deposition/erosion?

_____

_____

_____

_____

_____

_____

_____

_____

_____

 PHOTOCOPYING OF THIS PAGE IS RESTRICTED UNDER LAW.   ISBN: 9780170446907

5 Draw two annotated maps of the MCE that illustrate and explain this temporal variation in interacting natural processes.

6

Spatial variation of interacting natural processes at Muriwai Coastal Environment (MCE)

Spatial variation = how something varies (changes) within a space, between different locations.

In the Muriwai Coastal Environment (MCE) there are many different instances of where the dominant coastal processes vary spatially. This final chapter will closely analyse *three* of these variations and how they subsequently shape the natural environment and its features:

1 Spatial variation of the direction of longshore drift and how this changes the size of the beach in different areas.
2 Spatial variation of types of vegetation growth and how this interacts with the process of aeolian deposition at the dunes.
3 Spatial variation of water-layer weathering (salt crystallisation) on Fisherman's Rock.

Spatial variation 1 — direction of longshore drift

Key interacting processes:
Wave refraction *interacts with* longshore drift.

Fine Mitiwai sediment at the MCE is constantly moving along the coast. The natural process of longshore drift occurs on Muriwai Beach primarily in a northward direction. Waves can approach the coast at an angle from the southwest because of the direction of the southwesterly prevailing wind, blowing onshore from the Tasman Sea. The swash of the waves carries material up the beach at an angle. The backwash then flows back to the sea in a straight line at 90° to the shoreline. This repetition of swash and backwash transports material northwards in a zigzag motion. This process brings sediment into the MCE from the south and then keeps carrying it northwards. To sum up here, the predominant way that the process of longshore drift operates at Muriwai Beach is northward. *However, there is one significant spatial variation in this process.*

How does the process of longshore drift vary spatially in the MCE?

Despite this northward longshore drift of Mitiwai sediment at Muriwai Beach, there is one significant variation in this pattern. At the southern end of Muriwai Beach, along a 500 m stretch right next to Otakamiro Headland, longshore drift happens in the opposite direction, transporting sediment southwards towards the headland. This is because the process of longshore drift interacts with the wave refraction that is occurring at and around the headland. Wave refraction occurs at Otakamiro Headland because the direction in which waves approach from the southwest is altered by the shape of the MCE coastline. Otakamiro Headland protrudes out into the Tasman Sea by 500 m and the sea floor around it is significantly shallower than to the north and south. Waves travel faster in the deeper

PHOTOCOPYING OF THIS PAGE IS RESTRICTED UNDER LAW. ISBN: 9780170446907

water yet are slowed down as they approach the headland. In the shallower water the waves lose more energy due to friction so slow down, causing the wave to bend around the headland. Immediately north of the headland on Muriwai Beach (an approximately 500 m stretch), this causes waves to hit the Muriwai Beach coastline from a northwesterly angle. This in turn causes longshore drift to operate in a southward direction in this short portion of beach right next to the headland. The waves push sediment along in this different approach of the wave, and is a clear indication of how the process of wave refraction interacts with the process of longshore drift.

How does the spatial variation in longshore drift shape Muriwai Beach?

At Muriwai Beach it is clearly visible how the spatial variation of longshore drift shapes the coastline. For most of Muriwai Beach the predominant longshore drift direction of northward moves sediment along the beach, keeping it at this present time at a width of about 100 m.

At the southern 500 m stretch next to Otakamiro Headland the beach is significantly wider, approximately 200 m wide. This is because the southward-moving longshore drift (due to interaction with wave refraction) pushes large amounts of sediment southwards. Otakamiro Headland acts as a barrier to this southwards-moving sediment, trapping it and causing it to build up. As a result, the beach at the southern 500 m end of Muriwai Beach is significantly wider.

These variations in process and landforms are shown below in the annotated photograph.

Annotated photograph of Muriwai Beach showing spatial variation of longshore drift and consequent spatial variation in width of beach

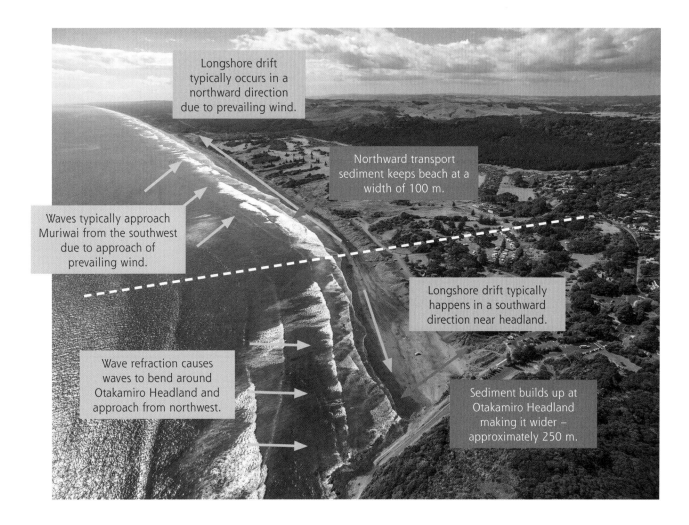

Annotated map of MCE showing spatial variation in direction of longshore drift due to interaction with process of wave refraction

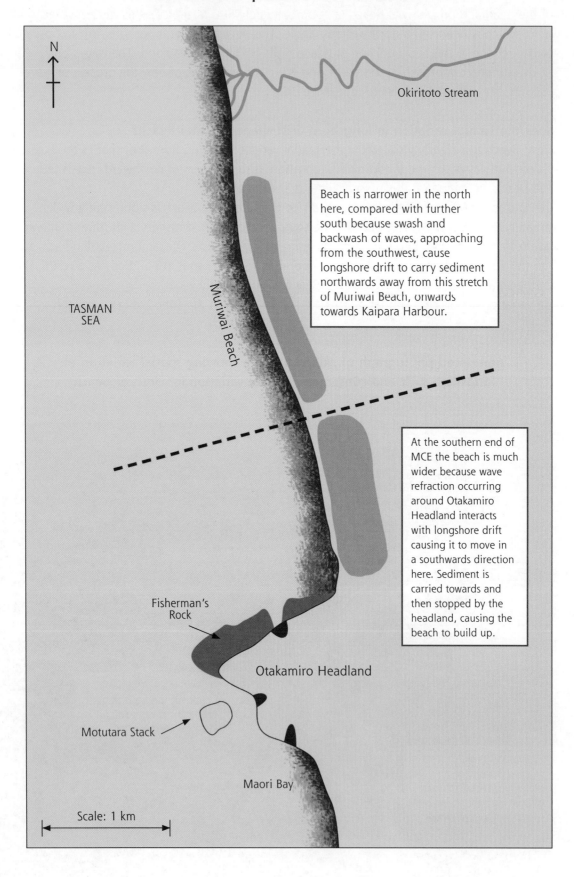

N

Okiritoto Stream

TASMAN
SEA

Muriwai Beach

Beach is narrower in the north here, compared with further south because swash and backwash of waves, approaching from the southwest, cause longshore drift to carry sediment northwards away from this stretch of Muriwai Beach, onwards towards Kaipara Harbour.

At the southern end of MCE the beach is much wider because wave refraction occurring around Otakamiro Headland interacts with longshore drift causing it to move in a southwards direction here. Sediment is carried towards and then stopped by the headland, causing the beach to build up.

Fisherman's Rock

Otakamiro Headland

Motutara Stack

Maori Bay

Scale: 1 km

     PHOTOCOPYING OF THIS PAGE IS RESTRICTED UNDER LAW.   ISBN: 9780170446907

Tasks

1 Explain why the general pattern of longshore drift at Muriwai Beach operates in a northward direction.

_____

_____

_____

_____

_____

_____

_____

_____

_____

_____

2 Explain why at the southern end of Muriwai Beach, longshore drift operates in a predominantly southward direction. Make sure you discuss interaction in this answer.

_____

_____

_____

_____

_____

_____

_____

_____

_____

_____

3 How does this spatial variation in the predominant direction of longshore drift shape Muriwai Beach?

_____

_____

_____

_____

_____

_____

_____

_____

_____

_____

_____

4 Draw a detailed annotated map of MCE showing spatial variation in longshore drift and how this shapes Muriwai Beach.

    PHOTOCOPYING OF THIS PAGE IS RESTRICTED UNDER LAW.   ISBN: 9780170446907

Spatial variation 2 — process of aeolian deposition at Muriwai Beach south and north

Key interacting processes:
Vegetation growth *interacts with* aeolian deposition.

A reminder of how dunes are shaped at Muriwai Beach:

1 At low tide, solar heating dries out sediment on the beach making it friable.
2 Onshore winds with a velocity over the fluid threshold velocity (5.47 mps) are able to deflate (erode) sediment from the foreshore.
3 Sediment saltates (bounces) up the beach face towards the backshore zone.
4 In the backshore zone, at the frontal dunes, vegetation growth (e.g. spinifex and marram) slows down wind flow resulting in aeolian deposition.
5 This deposition results in dune accretion, making the dunes taller and wider.

This process operates right along Muriwai Beach, as a result of the fast onshore winds, the plentiful supply of sediment and the expansive flat area of the backshore where vegetation allows aeolian deposition. However, when you look closely at how the dunes are shaped at Muriwai Beach, there is an obvious difference between the dunes at the southern 500 m stretch and the dunes further north – see photo below.

Annotated photograph showing spatial variation in height and shape of dunes at Muriwai Beach between southern sector and northern sector

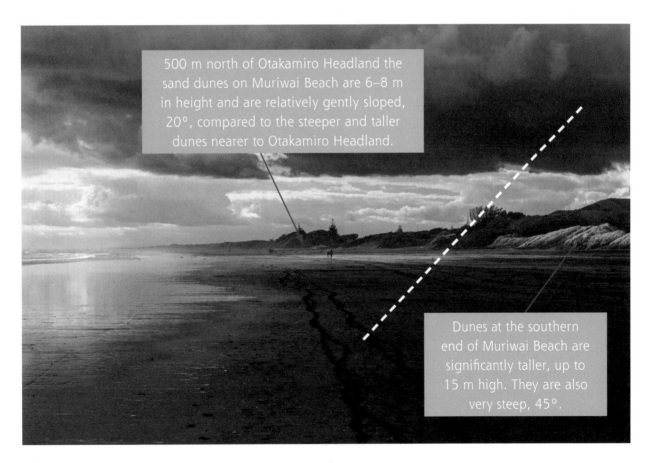

500 m north of Otakamiro Headland the sand dunes on Muriwai Beach are 6–8 m in height and are relatively gently sloped, 20°, compared to the steeper and taller dunes nearer to Otakamiro Headland.

Dunes at the southern end of Muriwai Beach are significantly taller, up to 15 m high. They are also very steep, 45°.

So why do the dunes vary in their height and slope between the north and south of Muriwai Beach? The answer lies in spatial variation of the process of vegetation growth and how this interacts with the process of aeolian deposition.

Northern Muriwai Beach (500 m north of Otakamiro Headland to Okiritoto Stream)

The dunes further north from Otakamiro Headland have been extensively planted with spinifex grasses. Spinifex grows in a unique way compared to other dune grasses. As sediment being transported falls around the spinifex and it becomes buried, the grass must grow out to survive. Most grasses grow their shoots (leaves) upward, however spinifex has long tendrils, also known as runners, that grow forwards and down the front of the dune (see photograph, right). This type of vegetation growth interacts with further aeolian deposition of sediment, influencing how it occurs. As the grass grows forward and down the front of the dune, it encourages sediment to be deposited by the wind in this area in front of the dune. As a result, the dune does not necessarily become taller and taller, yet instead dune accretion (build-up) occurs in front of the dune. The dune

Wide and gently sloped dunes in the northern section of MCE, vegetated with spinifex.

becomes wider, rather than taller, and much more gently sloped. The dunes here are approximately 6–8 m in height and have a slope of approximately 20°.

 PHOTOCOPYING OF THIS PAGE IS RESTRICTED UNDER LAW. ISBN: 9780170446907

Southern Muriwai Beach (500 m stretch next to Otakamiro Headland)

The dunes in this southern section of the beach vary significantly compared to their counterparts further north. Again, this is the result of the interaction between the type of vegetation growth and aeolian deposition of sediment being transported by the strong onshore winds. At these foredunes, there is very little spinifex growing — instead these dunes are dominated by an Australian grass called marram, which was planted in the 1950s. When marram was planted, it was hoped that it would stabilise the dunes and prevent their erosion by binding the sediment and increasing aeolian deposition. While this vegetation growth has helped to increase aeolian deposition, it happens in a very different way from the dunes to the north that are covered in spinifex. The marram grass grows in an upwards direction (as opposed to the forward direction of growth seen with spinifex). As sediment is deposited around the grass, smothering it, the grass grows upwards. This continued process of deposition and vegetation growth causes dune accretion to occur in an upwards fashion. The result is that dunes become taller — the dunes are up to 12 m here, compared with the 6–8 m dunes further north. It also means that the dunes are much steeper, at 45°. Waves also interact here — the waves can easily undercut the front of the steep dunes, eroding them out, resulting in slumping at the front of the dune. Wave erosion at these dunes is increased because of their shape.

Significantly taller dunes at the southern end of MCE resulting from the upwards growth of marram.

Annotated map of MCE showing variation in dune shape due to variation in interaction of aeolian and vegetation processes

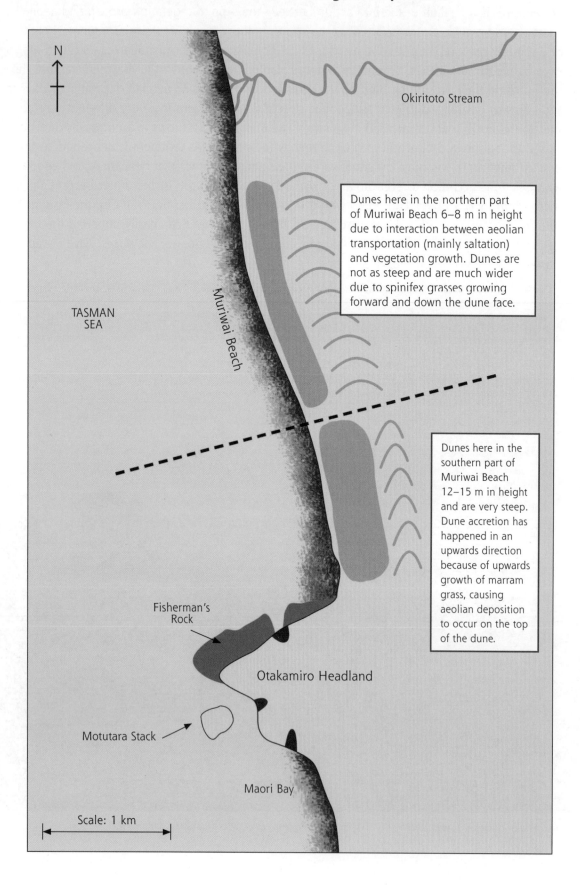

N

Okiritoto Stream

TASMAN SEA

Muriwai Beach

Dunes here in the northern part of Muriwai Beach 6–8 m in height due to interaction between aeolian transportation (mainly saltation) and vegetation growth. Dunes are not as steep and are much wider due to spinifex grasses growing forward and down the dune face.

Dunes here in the southern part of Muriwai Beach 12–15 m in height and are very steep. Dune accretion has happened in an upwards direction because of upwards growth of marram grass, causing aeolian deposition to occur on the top of the dune.

Fisherman's Rock

Otakamiro Headland

Motutara Stack

Maori Bay

Scale: 1 km

    PHOTOCOPYING OF THIS PAGE IS RESTRICTED UNDER LAW.   ISBN: 9780170446907

Tasks

1 How do the processes of vegetation growth interact with the process of aeolian deposition?

_____

_____

_____

_____

_____

_____

_____

_____

_____

_____

_____

2 Compare spatial variation in vegetation growth on the foredunes between the northern and southern sectors of Muriwai Beach at the MCE. Refer to specific types of vegetation and how they grow.

_____

_____

_____

_____

_____

_____

_____

_____

_____

_____

_____

3 How do these different types of interactions result in spatial variation in the shape of the dunes at Muriwai Beach?

_____

_____

_____

_____

_____

_____

_____

_____

_____

_____

4 Draw a detailed annotated map of MCE showing spatial variation in vegetation growth and aeolian deposition, and how this shapes the foredunes at Muriwai Beach.

PHOTOCOPYING OF THIS PAGE IS RESTRICTED UNDER LAW.
ISBN: 9780170446907

Spatial variation 3 — variation in rates of water-layer weathering on Fisherman's Rock shore

Key interacting processes: Wetting and drying *interacts with* salt crystallisation.

Shore platform

Fisherman's Rock is the large shore platform surrounding the northern side and front of Otakamiro Headland. It has formed at the high-water mark due to 10,000 years of cliff retreat. Today the platform is being weathered on a daily basis by a type of salt-crystallisation weathering known as water-layer weathering.

Water-layer weathering occurs because the platform is submerged twice daily by the high tide. When the tide recedes and exposes the platform to the atmosphere, a layer of sea-water remains on it. When the sun heats this water layer it evaporates the water, leaving behind salt crystals. The salt crystals form in the rock matrix of the platform and as the water fully evaporates, the crystals grow in size. This crystal growth puts minuscule amounts of pressure on the structure of the rock on the surface of the platform, which in turn weakens it and breaks off small particles of rock from time to time. Over thousands of years this has lowered the platform by up to 100 cm. It has also made the platform flatter, and widened out joints in the rock.

A significant spatial variation to this process occurs at the very front edge of the Fisherman's Rock shore platform. This front edge is known as the 'rampart' and water-layer weathering occurs much slower here than the rest of the platform. The result is that the platform is raised around this rampart, by up to 20 cm higher than the rest of this platform. The reason for this spatial variation is because waves constantly splash this part of the platform, not allowing it to dry out as frequently as the rest of the platform. Therefore, salt crystals will not form as often. This is a good example of interaction of processes — higher rates of wetting and lower rates of drying slow down water-layer weathering at this part of the headland.

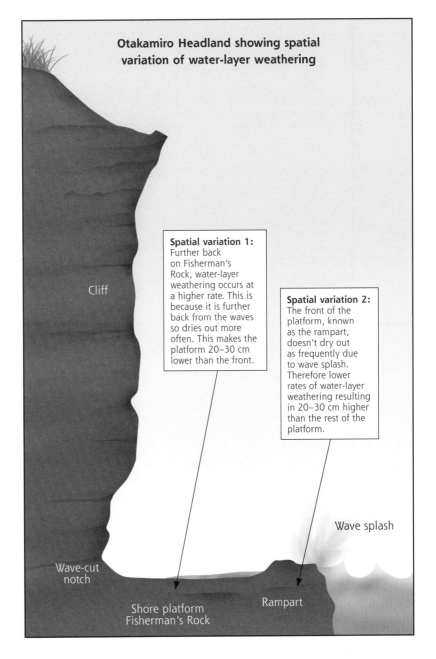

Otakamiro Headland showing spatial variation of water-layer weathering

Spatial variation 1: Further back on Fisherman's Rock, water-layer weathering occurs at a higher rate. This is because it is further back from the waves so dries out more often. This makes the platform 20–30 cm lower than the front.

Spatial variation 2: The front of the platform, known as the rampart, doesn't dry out as frequently due to wave splash. Therefore lower rates of water-layer weathering resulting in 20–30 cm higher than the rest of the platform.

Cliff

Wave-cut notch

Shore platform Fisherman's Rock

Rampart

Wave splash

Task

On the blank cross-section of Otakamiro Headland (below), construct annotations that fully explain how interacting natural processes have resulted in spatial variation of features on the wave-cut platform.

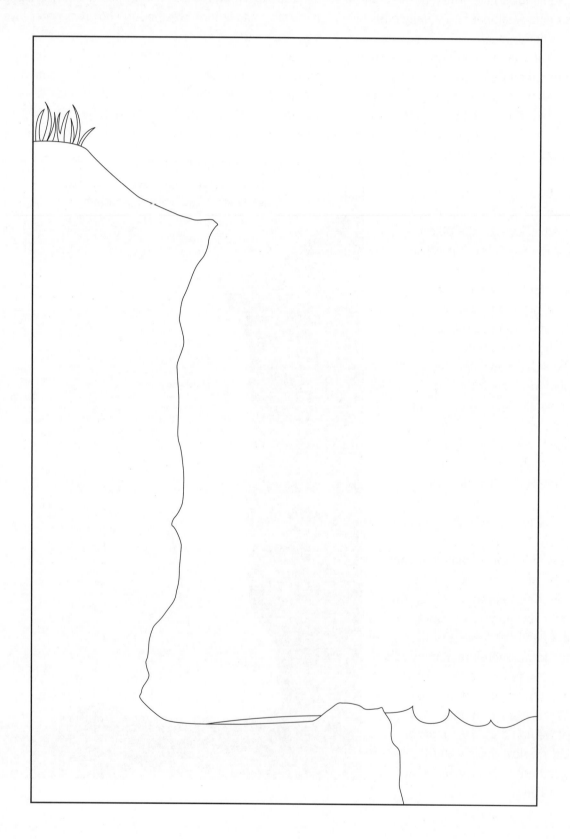

 PHOTOCOPYING OF THIS PAGE IS RESTRICTED UNDER LAW. ISBN: 9780170446907

Answers

Chapter 1 (p. 7)

1 A coast is an area where land, sea and the atmosphere meet and interact.

2 No, coasts vary significantly. Some differences to consider:
   - cliffed vs beached coastlines
   - constructive vs destructive coastlines.

3 *Teacher to mark.* Student has drawn a diagram based on example on page 6. Student has provided annotations describing the backshore, foreshore, inshore and offshore.

4 Constructive coastlines have gentler waves. Higher rates of deposition to erosion build up sediment on the beach.
   Destructive coastlines experience powerful, high-energy waves that erode the coastline.

Chapter 2 (pp. 10–12)

1 A wave is simply the transfer of kinetic energy through the surface of a body of water.

2 The vast majority of ocean waves are generated through wind blowing over the surface of the ocean. Factors that can influence the size and energy of waves include: velocity of wind; duration of wind; fetch (the distance a wave travels).

3 *Teacher to mark.* Student has drawn a diagram based on example on page 8. Student has correctly labelled key parts.

4 Waves in the open ocean are called 'swells' — the energy is moving forward through the surface of the ocean, but not the water itself. A person swimming here would simply bob around in an orbital motion.
   Waves at the coastline form breakers, the sea floor interacts with the base of the wave, slowing it down. The top of the wave maintains a higher speed than the base and topples over.

5 Swash: the water of a broken wave that washes up a beach face.
   Backwash: excess swash returning back down the beach face towards the sea as a result of the effect of gravity.

6 *Teacher to mark.* Student has drawn a diagram based on examples on page 9. Student has labelled and annotated.

7 *Teacher to mark.* Wave refraction is the bending of waves around an object (e.g. headland) due to changes to the shape of the seafloor. Student has drawn a diagram based on example on page 10. Student has labelled and annotated.

(pp. 15–18)

1 Erosion is the wearing away/breaking up/destruction of land/rock/sediment.

2 Wave pounding — force of the wave hammering away at rock or cliff.
   Wave hydraulic pressure — breaking apart of rock through waves pressurising air into cavities/fissures/joints.
   Abrasion — also known as corrasion. Scraping effects as sand/pebbles/shingles are thrown against coastline.
   Attrition — loose sediment in the surf colliding or rubbing together.
   Solution — also known as corrosion. Chemical reactions result in breakdown of geology.

3 *Teacher to mark.* Student has drawn separate diagrams based on example on page 13. Student has annotated clearly.

4 Erosivity — force of the waves.
   Erodibility — strength (or weakness) of geology of coastline.

5 *Teacher to mark.* Student has drawn on ideas discussed on page 14.

(pp. 20–21)

1 Breakdown of coast/cliffs/rocks that has resulted from process not directly involving waves. Student has discussed two examples: mass movement and salt crystallisation.

2 Ideas could include: does not directly involve waves; may result from atmospheric processes or the effects of gravity; sub-aerial process may occur at slower rates.

3 One process affects another process by speeding up/slowing down/stopping/modifying the direction it occurs, etc.

4 First diagram: High rates of sub-aerial weathering will weaken the geological structure of the coastline — this, in turn, will increase rates of marine erosion processes such as wave hydraulic pressure and abrasion.
Second diagram: High rates of marine erosion may weaken cliffs and even expose new rocks. Undercutting of a cliff through marine erosion will weaken the upper part of the cliff — this will result in higher rates of mass movement as gravity takes effect.

(pp. 24–25)

1 Wave-cut notch — an eroded indentation found running along the base of the cliff between the high- and low-water marks.
Shore platform — flat rock at base of a cliff formed by cliff retreat.
Arch — a natural tunnel-like formation running through one side of a headland to the other.
Stack — where an arch has collapsed, leaving a large tower of rock in front of the main headland.
Stump — this is the base of the stack that remains after erosion has caused the stack to collapse.

2 *Teacher to mark.* Student has explained how marine erosion leads to cliff retreat and forms a shore platform.

3 *Teacher to mark.* Student has drawn diagram(s) based on examples on pages 23 and 24. Student has annotated clearly.

(pp. 28–29)

1 *Teacher to mark.* Student has provided descriptions and diagrams based on the information on page 26.

2 Discusses points such as these:

Littoral drift
- Happens in the nearshore zone (surf zone) in the water.
- Sediment is carried by the longshore current by processes such as suspension, saltation, traction.

Longshore drift
- Happens on the beach in the swash zone.
- Swash carries sediment up the beach at an angle due to oblique approach of waves.
- Backwash makes sediment roll back down the beach into the surf.
- This process repeats, causing sediment to zigzag its way along the beach.

3 *Teacher to mark.* Student has drawn detailed and annotated diagrams based on the information on pages 26 and 27.

(p. 31)

1 Summer beaches have a steeper gradient due to constructive waves depositing sediment on the beach. Offshore bars are smaller during summer.
Winter beaches have a flatter gradient due to destructive waves eroding sediment from the beach. Sediment is stored offshore, making offshore bars larger.

2 *Teacher to mark.* Student effectively used the geographic terminology of 'interaction' to explain how summer and winter climatic processes cause waves to be constructive or destructive.

(pp. 37–39)

1 Sand dunes are ridges of sand found running along the upper part of a sandy beach's backshore zone. They provide a natural barrier to beach erosion and are usually vegetated with dune plants such as grasses.

2 *Teacher to mark.* Student has identified four different conditions, including: onshore wind direction, high-enough wind velocity, low tide, solar heating.

3 Saltation — sediment travels short distances through the air in a skipping/bouncing motion.
Suspension — with higher wind gusts, fine sediment particles float through the air for long distances.
Surface creep — sediment particles roll or slide along the surface of the beach.

   PHOTOCOPYING OF THIS PAGE IS RESTRICTED UNDER LAW.   ISBN: 9780170446907

4 *Teacher to mark.* Student effectively used the geographic terminology of 'interaction' to explain how increased vegetation growth results in higher rates of aeolian deposition. This results in dune accretion.

5 Human use/abuse of frontal dunes may result in vegetation removal and reduced rates of vegetation growth. This interacts with aeolian processes — lower rates of aeolian deposition and higher rates of aeolian erosion on the dunes.

6 *Teacher to mark.* Student has discussed several differences and similarities.

Chapter 3 (p. 43)

1 *Teacher to mark.* Student has drawn a detailed and annotated diagram based on example on page 41.

(pp. 52–53)

1 2 km
2 50–100 m, 0–5°
3 It has a dissipative (winter) profile for most of the year.
4 0.0025 mm
5 Titanomagnetite
6 Black
7 Taranaki, Ruapehu, Tongariro, Ngauruhoe
8 Longshore drift, littoral drift
9 Okiritoto Stream, Otakamiro Headland
10 5–12 m
11 Spinifex, marram, pikao
12 They help to protect human assets from coastal erosion.
13 30 m
14 Volcanism
15 Faulting
16 Dense, erosion-resistant rock
17 Manukau Breccia
18 Fisherman's Rock
19 Motutara Stack
20 Maori Bay
21 Cliffs
22 Pillow lava

Chapter 4 (pp. 56–57)

1 Discusses points including:
 • littoral transportation brings sediment into the Muriwai Coastal Environment
 • supplies sediment to the process of wave transportation/deposition

 • this is an example of processes interacting
 • increased rates of marine deposition at Muriwai Beach.

2 Discusses points including:
 • exposed setting on West Coast
 • high wind speeds
 • prevailing winds are predominantly onshore
 • flat beach gradient/wide beach/large inter-tidal zone
 • case study specific detail.

3 *Teacher to mark.* Student has drawn a detailed and annotated diagram based on example on page 55.

(pp. 61–63)

1 *Teacher to mark.* Student has drawn a series of three appropriate detailed and annotated diagrams based on eight diagrams on page 58.

2 *Teacher to mark.* Student has drawn a detailed and annotated diagram based on examples on pages 60 and 61.

Chapter 5 (pp. 67–69)

1 Discusses points including:
 • climatic processes in winter involve a high occurrence of low-pressure systems resulting in higher wind velocities
 • climatic processes in summer involve a high occurrence of high-pressure systems resulting in lower wind velocities.

2 Discusses points including:
 • differences in swash/backwash ratios between summer and winter
 • winter has higher rates of erosion; summer has higher rates of deposition
 • this is due to interaction between waves processes and climatic processes
 • specific detail related to Muriwai Coastal Environment.

3 BEACH:
 • Flatter in winter and fractionally steeper in summer.
 • Links this to interacting processes.
 • Student has included specific detail related to Muriwai Coastal Environment.
 DUNES:
 • Face of frontal dunes more gently sloped in summer; heavily undercut and steep in winter.
 • Links this to interacting processes.
 • Student has included specific detail related to Muriwai Coastal Environment.

4 *Teacher to mark.* Student has drawn detailed and annotated diagram(s) based on examples on page 66.

(pp. 73–74)

1 See table below.

2 *Teacher to mark.* Student has drawn detailed and annotated diagrams based on examples on page 71.

(pp. 78–79)

1 A slug is a large body of sediment travelling up a coastline due to processes of littoral transportation. On New Zealand's west coast, it moves northwards very slowly, taking hundreds to thousands of years. This slug is held up at times by geographic features such as harbours and river mouths.

2 Resulted in growth of the beach due to higher rates of wave deposition. Includes specific detail.

3 Resulted in retreat of the beach due to higher rates of wave erosion. Includes specific detail.

4 Higher rates of littoral transportation results in higher rates of wave deposition and vice versa.

5 *Teacher to mark.* Student has drawn detailed and annotated diagrams based on examples on page 77.

Chapter 6 (pp. 83–84)

1 Results from southwesterly prevailing wind, due to Coriolis effect.

2 Interaction between wave refraction at Otakamiro Headland and longshore drift.

3 Beach is wider at southern end and narrower further north. Student has explained this pattern and used specific detail.

4 *Teacher to mark.* Student has drawn a detailed and annotated diagram based on example on page 82.

(pp. 89–90)

1 Grasses that grow in a forward direction down the face of the dune result in deposition at the front of the dune. Grasses that grow in an upward direction result in deposition occurring on top of the dune.

2 At the southern end, grasses such as marram and pikao grow in an upward direction. Further north, spinifex grows in a forwards direction down the face of the dune.

3 Dunes at the southern end have grown tall and steep, whereas further north they are wider and more gently sloped. Student has explained how this is resulting from interaction of processes.

4 *Teacher to mark.* Student has drawn a detailed and annotated diagram based on example on page 88.

(p. 92)

1 *Teacher to mark.* Student has drawn a detailed and annotated diagram based on example on page 91.

	1950	Present day
Explain how the process of vegetation growth was operating at MCE dunes.	Vegetation growth was occurring at very slow rates.	Vegetation growth is occurring at much higher rates. Vegetation cover on the dunes is much greater today.
How was this a direct result of human management/ mismanagement?	Vegetation was being removed to make way for farming and other activities.	Vegetation is being actively planted, managed and protected in order to stabilise the dunes.
How did the process of vegetation growth during this period interact with aeolian processes?	Increased aeolian erosion (deflation) of the dunes at Muriwai Beach.	Increased rates of aeolian deposition on the dunes.
How did this shape the environment, in particular the frontal dunes?	Dune system heavily degraded. Large amounts of sediment transported inland.	Dune system has been largely stabilised. Dunes have been able to grow in height and width.

 PHOTOCOPYING OF THIS PAGE IS RESTRICTED UNDER LAW. ISBN: 9780170446907